CAUSAL REASONING IN PHYSICS

Much has been written on the role of causal notions and causal reasoning in the so-called "special sciences" and in common sense. But does causal reasoning also play a role in physics? Mathias Frisch argues that, contrary to what influential philosophical arguments purport to show, the answer is yes. Time-asymmetric causal structures are as integral a part of the representational toolkit of physics as a theory's dynamical equations. Frisch develops his argument partly through a critique of anti-causal arguments and partly through a detailed examination of actual examples of causal notions in physics, including causal principles invoked in linear response theory and in representations of radiation phenomena. Offering a new perspective on the nature of scientific theories and causal reasoning, this book will be of interest to professional philosophers, graduate students, and anyone interested in the role of causal thinking in science.

MATHIAS FRISCH is Professor of Philosophy at the University of Maryland, College Park.

CAUSAL REASONING IN PHYSICS

MATHIAS FRISCH

CAMBRIDGE
UNIVERSITY PRESS

CAMBRIDGE
UNIVERSITY PRESS

University Printing House, Cambridge CB2 8BS, United Kingdom

One Liberty Plaza, 20th Floor, New York, NY 10006, USA

477 Williamstown Road, Port Melbourne, VIC 3207, Australia

4843/24, 2nd Floor, Ansari Road, Daryaganj, Delhi - 110002, India

79 Anson Road, #06-04/06, Singapore 079906

Cambridge University Press is part of the University of Cambridge.

It furthers the University's mission by disseminating knowledge in the pursuit of
education, learning and research at the highest international levels of excellence.

www.cambridge.org
Information on this title: www.cambridge.org/9781316649657

First published 2014
First paperback edition 2017

A catalogue record for this publication is available from the British Library

ISBN 978-1-107-03149-4 Hardback
ISBN 978-1-316-64965-7 Paperback

Contents

Figures

Acknowledgments

Early research on the issues examined in this book was funded by the US National Science Foundation (award number SES-0646677), for which I am grateful.

Large parts of this book were written while I was a visitor at the Munich Center of Mathematical Philosophy (MCMP) at the Ludwig Maximilians University. During the academic year 2011–12 I had a fellowship for experienced researchers awarded by the Alexander-von-Humboldt Foundation. During the year 2012–13 I was a senior visiting fellow at the MCMP. I am grateful to the Humboldt Foundation for their generous support. I thank my host during the Humboldt fellowship, Ulises Moulines, and the MCMP's co-directors Hannes Leitgeb and Stephan Hartmann for their support and hospitality. My work on the Ritz-Einstein debate, examined in Chapter 7, benefited from many conversations with Wolfgang Pietsch from the Technical University at Munich, and some sections of the chapter are taken from joint work with Wolfgang.

Some chapters of the book are drawn from previously published work:
Chapter 2 is based on "Users, Structures, and Representation,"
The British Journal for the Philosophy of Science 2014; doi: 10.1093/bjps/axt032

Chapter 3 includes sections from: "Physics and the Human Face of Causation," *Topoi. An International Review of Philosophy* (guest editors Federica Russo and Phyllis Illari). doi 10.1007/s11245-013-9172-0

Chapter 4 includes sections from: "Causal Models and the Asymmetry of State Preparation" in EPSA Philosophical Issues in the Sciences. Launch of the European Philosophy of Science Association, M. Suárez, M. Dorato, and M. Rédei, eds. Springer, Berlin (2010).

Chapter 5 includes sections from: "No Place for Causes? Causal Skepticism in Physics," *European Journal for Philosophy of Science* 2:3 (October 2012), 313–36.

Chapter 6 incorporates the following two papers: "'The Most Sacred Tenet'? Causal Reasoning in Physics," *British Journal for the Philosophy of Science* 60 (September 2009), 459–74. And: "Causality and Dispersion: A Reply to John Norton," *British Journal for the Philosophy of Science* 60 (September 2009), 487–95.

Chapter 8 includes large parts from: "Does a Low-Entropy Constraint Prevent Us from Influencing the Past?," in *Time, Chance, and Reduction*, edited by Andreas Hüttemann and Gerhard Ernst, Cambridge, UK: Cambridge University Press (2010), 13–33.

CHAPTER I

Introduction

1. From Helmholtz to Russell: a very brief historical sketch

Is there a place for causal reasoning in physics? Many readers might think that the answer to this question must obviously be "yes." Since it is the aim of science to explain the natural world, one might argue, and since the search for explanations is just a search for causes, causal reasoning obviously plays an important role in physics. Since physics is arguably the most fundamental science, it must be concerned with discovering the most fundamental causal relations.

Indeed, that physics is concerned with the search for causes appears to have been a widely held view in the late eighteenth century and up until the middle of the nineteenth century. For example, the German physicist Hermann von Helmholtz (1821–1894), in a public lecture, characterized the aim of physics as follows:

> Our demand to understand natural phenomena, that is, to discover their laws, is a different way of expressing the demand that we are to search for the forces that are the causes of the phenomena. The lawfulness of nature is conceived of as causal relationship, as soon as we recognize nature's independence from our thought and from our will. Thus when we ask about the progress of science as a whole, we will have to judge it according to the extent in which the recognition and the knowledge of causal connections, encompassing all natural phenomena, have progressed. (Helmholtz, 1896, 40, my translation)

For Helmholtz the centrality of causes is underwritten by a conception of forces as causes of motion. In his talk Helmholtz attempted to develop a unified conception of science with mechanics at its foundation:

> If motion is the primary change, which forms the basis of all other changes in the world, then all elementary forces are forces of motion; and the ultimate

aim of science is to find those motions and their forces that form the basis of all other changes – that is, for science to dissolve into mechanics. (Helmholtz, 379)

Thus, for Helmholtz the ultimate aim of science is to find the basic forces, and these forces are understood as causes of fundamental motions. Indeed, a conception of forces as causes of motion appears to have been widely endorsed up until the middle of the nineteenth century. The physicist Gustav Theodor Fechner (1801–1897), for example, puts this view as follows: "that physicists often speak of force simply as the cause of motion" (Fechner 1864, 126, my translation).

Yet the view of physics as a search for causes was increasingly questioned in the last decades of the nineteenth century, and there now exists a long and distinguished tradition denying that causal notions can play a legitimate role in physics.[1] In the introduction to his *Vorlesungen zur Mechanik* (Lectures on Mechanics), Gustav Kirchhoff (1824–1887) criticizes the definition of forces as the causes of motion (and the very conception of science as the search for the basic forces as causes championed by Helmholtz) as being infected by unacceptable vagueness:

> It is customary to define mechanics as the science of *forces* and to define forces as the *causes* that produce motion or *strive* to produce motion . . . [but this definition] is infected by the vagueness from which the notions of cause of striving cannot be freed . . . Given the precision that otherwise characterizes inferences in mechanics, it appears to be desirable to remove such obscurities even if this were possible only through a restriction of its purpose. For this reason I take the task of the science of mechanics to be to describe the motions found in nature, and to describe them completely and as simply as possible. By this I mean that the aim is to state *what* the phenomena are which occur, rather than to determine their *causes*. (Kirchhoff, 1876, p. v, my translation; italics in the original)

The term "force" still plays a role in Kirchhoff's treatment, but forces are defined implicitly through the equations of mechanics: "In order to remove any obscurity it is sufficient to define the notion of forces only insofar as every theorem in mechanics which speaks of forces can be translated into equations" (*ibid.*, p. vi).

That the concept of cause is inherently and irredeemably vague is a criticism that has often been repeated since. In the early twenty-first century

---

[1] For a more detailed and excellent discussion (in German) of the history of the role of causal notions in physics in the nineteenth century, see Hüttemann (2013). My brief survey here follows Hüttemann's discussion in broad outline.

we find this view defended, for example, by the philosophers of physics John Earman and John Norton. Earman derides appeals to causal notions in physics by maintaining that the contest of conflicting intuitions about causal notions "may generate many learned philosophical articles," but that "a putative fundamental law of physics must be stated as a mathematical relation without the use of words that require a PhD in philosophy to apply (and two other PhDs to referee the application, and a third to break the tie in the inevitable disagreement of the first two)" (Earman 2011, 494). He insists that explanations in physics may not involve any causal "philosophy-speak" (Earman 2011, 494). Norton expresses a similar view, claiming that "the conditions of applicability [of causal notions] are obscure" (Norton 2009, 481).

Whereas Kirchhoff's criticism of causal notions appears to be directed against a specific conception of cause as that which "produces" or "brings about" its effects (or "strives" to bring about its effects), later criticisms are directed against what appear to be less metaphysically loaded conceptions of "cause" as well. Thus, although the physicist Ernst Mach initially adopted John Stuart Mill's Humean regularity account of causation, according to which "the law of Causation . . . is but the familiar truth that invariability of succession is found by observation to obtain between every fact in nature and some other fact which has preceded it" (Mill 1875, III-v-§2), Mach later rejected the account and argued for a complete rejection of causal notions in physics as follows:

> When we speak of cause and effect, then we arbitrarily emphasize those aspects, the connections among which are the ones on which we have to focus, when we represent a fact from a certain perspective that is important to us. In nature there are no cause nor an effect. Nature exists only once. Repetitions of the same cases, in which A would always be linked with B, thus same effects under the same circumstances, thus the essence of the connection between cause and effect, exist only in the abstraction, which we undertake in order to represent the facts. (Mach 1901, 4.4.3, p. 513)

Mach here argues that causal regularities of the form "All A's are followed by B's" are the result of abstracting from the multitude of factors on which the occurrence of an event depends. The argument can be fleshed out in a bit more detail as follows. Imagine we are interested in representing the motion of a particular billiard ball B on a billiard table. In providing a mathematical model of the ball's motion, it may be useful to focus only on the motion of the cue ball and its collision with B as *the* "cause" of B's motion and to abstract from the dependence of the ball's motion on any other factors, such as the gravitational forces exerted by the billiard players, nearby physical objects, or the sun. In many contexts it is appropriate to represent the motion of the ball in terms of a simple model that contains only the table and the balls on it, treats the collisions among balls as fully elastic and ignores gravitational forces. For a simple model, such as this, there will be regularities of the form "every ball at rest of mass m that is struck head-on by another ball with momentum p will move at velocity v." In principle, however, the motion of the ball depends on many other factors as well, which are ignored in the simple model. If we were to include these in our description of the collision event, we would find that the very complicated precise combination of factors on which the precise motion of ball B on a given occasion depends occurs exactly once. As Mach puts it, "nature exists only once." Thus, since the regularity "Whenever the full set of factors F occurs, they are followed by B" is instantiated only once, it is trivially true: corresponding to every true and complete description of the state of a system (or the world) at a time, and an event immediately succeeding that time, there is a true universal generalization of the form "The full set of factors F is followed by B."

For Mach this argument entails a complete elimination of the notion of cause: "If we aim to remove the traces of fetishism that are still attached to the notion of cause and if we realize that a cause can generally not be specified, but that a fact usually is determined by a whole systems of conditions, then this leads us to giving up the notion of cause completely" (Mach 1900, 433). The view that causal notions are in some sense perspectival, playing a role only in our representations ("Nachbildungen") of the world, and hence are not a legitimate part of physics, is a view that is also prominent among critics of causal notions in the late twentieth and early twenty-first centuries, as we will see in detail in subsequent chapters.

Mach's conclusion is that, in the advanced sciences, the concept of cause has been replaced by that of functional dependency:

> In the higher developed sciences the use of the concepts of cause and effect is more and more restricted and increasingly rare. The reason is that these concepts characterize a state of affairs only in a preliminary and incomplete manner and that they lack precision . . . As soon as one succeeds in characterizing the elements of events through measurable quantities [. . .], the dependencies among these elements can be represented much more completely and more precisely with the help of the concept of a function than through the indeterminate concepts of cause and effect. (Mach 1905, 278)

A functional dependency expresses the values some physical quantity (the "output") can take in terms of the values of other quantities (the "input").

The relation between the different quantities is a function, exactly if for each set of input values there corresponds exactly one output value. One prima facie advantage of expressing the relation between quantities in terms of functional dependencies among variables used to represent these quantities is that this appears to avoid the problem of trivialization: whereas the precise combination of values for the different input variables determining the value of the output variable may occur only once (and hence the corresponding causal regularity is trivially universally instantiated), the functional dependency relating input and output variables may be multiply instantiated.

Like Kirchhoff before him, Mach criticizes causal notions as being inherently vague. However, Mach's criticism of causal notions in physics and the sciences is more general and does not merely amount to a positivist or empiricist criticism of an overly metaphysical notion of causal "production" or of "bringing about." For Mach, even an empiricist Humean regularity notion of cause cannot be part of physics proper and has been replaced there by the more appropriate and more precise concept of functional dependency.

The English philosopher Bertrand Russell repeated many of Mach's criticisms in his famous and influential essay "On the Notion of Cause" (1912–13), which, at least in the English-speaking world, is much better known than Mach's earlier critique. Russell repeats both Mach's vagueness charge and the claim that even a regularity notion of causation is problematic, since true regularities would be instantiated at most once:

> The principle "same cause, same effect," which philosophers imagine to be vital to science, is therefore utterly otiose. As soon as the antecedents have been given sufficiently fully to enable the consequent to be calculated with some exactitude, the antecedents have become so complicated that it is very unlikely they will ever recur. (Russell 1912, 9)

Russell concludes, again following Mach, that the concept of cause in the advanced sciences has been replaced by the notion of functional dependency.

Russell's essay contains one additional criticism – a criticism that has often been repeated since. He points out that the notion of cause is time-asymmetric – effects do not precede their causes – whereas the laws of the basic theories of physics are time-symmetric: "the future 'determines' the past in exactly the same sense in which the past 'determines' the future" (Russell 1912, 15). From this contrast he concludes that physics is incompatible with causal notions. Appeals to the time symmetry of the dynamical equations constitute perhaps the predominant reason for why philosophers have, often without further argument, concluded that causal notions can play no role in physics. For example, the German philosopher of physics Erhard Scheibe maintains, after pointing to the contrast between time-symmetric laws and time-asymmetric causal relations, that "this suffices to seal the fate of event-causality" – of causation as a relation between pairs of events or event types (Scheibe 2006).

The overall lesson Russell draws from his discussion of causation in physics is that causal notions should be rejected *in general* as having no useful role in our conception of the world. In an oft-quoted passage, he says: "The law of causality, I believe, like much that passes muster among philosophers, is a relic of a bygone age, surviving, like the monarchy, only because it is erroneously supposed not to do harm" (Russell 1912, 1).

Although Russell's criticism of causal notions has received a lot of attention, in particular in the early years of the twenty-first century, the fact that he himself came to change his mind on the role of causal notions in science is much less frequently discussed. I will discuss Russell's later view briefly below (see Russell 1921; 1948; 1954). Yet not just Russell changed his mind; philosophy more generally seems to have moved away from a wholesale rejection of causal concepts of the kind that may have been fashionable during the heyday of logical positivism. Indeed, the position of many causal critics a century after the publication of Russell's essay seems to be closer to Mach's view than to the view Russell argued for in "On the Notion of Cause," since, unlike Russell, Mach appears to have allowed that causal relations can be a legitimate aspect of our partial and abstract representations of the phenomena. That causal notions can play a role in our representations of the phenomena from a particular perspective and in a particular context is a view with prominent defenders in the twenty-first century. The philosopher James Woodward, for example, argues in the essay "Causation with a Human Face" (Woodward 2007) – a paper we will discuss in detail in subsequent chapters – that it is precisely the fact that causation has a "human face" which constitutes the reason why causal notions do not sit well with our more fundamental theories of physics. Mach, however, as we saw, ultimately concluded from the per-spectival character of causal notions that such notions ought to play no role in the more highly developed sciences, whereas Woodward and others want to draw a distinction between the "special sciences" on the one hand, in which causal reasoning is thought to play an important role, and physics on the other, which does not allow for a legitimate place for causal concepts.

I have mentioned a number of anti-causal claims that have been promi-nent in discussions of causal notions in physics from the nineteenth century onward. Among these claims are the following: (i) the notions of cause and effect are inherently vague; (ii) this vagueness infects especially metaphysi-cally rich notions of causal production; (iii) a regularity account of causation is problematic, since the set of factors on which a given effect depends is so large that the true causal regularities would be instantiated at most once; (iv) causal notions are part only of our abstract representations of the phe-nomena, and hence may be thought to be context- and interest-relative; and (v) the notion of cause is time-asymmetric, whereas the dynamical laws of the fundamental or established theories of physics are time-symmetric. In the following chapters I discuss these claims and several others to argue that they cannot be fashioned into arguments that succeed in showing that causal reasoning has no legitimate role to play in physics.

2. Distinct philosophical projects

The question whether there is a place for causal notions in physics can be asked within the context of several different philosophical projects.[2] The first such project is a metaphysical project interested in determining the metaphysical "grounds" for causal claims. The main division in the meta-physics of causation is between defenders of broadly Humean accounts and defenders of accounts that are broadly non-Humean. Humeans fol-low the Scottish philosopher David Hume (or at least follow Hume as he has traditionally been understood) in rejecting fundamental modali-ties. According to Humeans, the universe fundamentally is composed of a distribution of categorical properties and relations instantiated by funda-mental entities throughout spacetime. This distribution is often referred to as "the Humean mosaic." For the Humean, all modal claims, including causal claims, are made true by features of the mosaic, such as regularities in the distribution of categorical properties. By contrast, non-Humeans believe that modal properties, such as necessitation relations or disposi-tional essences, are themselves fundamental properties. For example, one might hold that it is in the nature of objects with mass to attract other massive objects. Or one might hold that causal laws are a fundamental feature of reality and that it is in virtue of such laws that earlier states of the world produce or bring about later states.

---

[2] The distinctions I am drawing here are similar to ones Woodward drew in his Presidential Address at the 2012 Meeting of the Philosophy of Science Association in San Diego, CA.

Within the context of the metaphysical project, the question concerning the place of causal notions in physics becomes the question whether a certain metaphysical account of causation *receives support from*, *is at least compatible with*, or *is undermined* by certain central features of physics. The philosopher Tim Maudlin, for example, argues that the laws of physics are fundamentally causal laws governing how earlier states generate later states of a system (Maudlin 2007). Also, the philosopher Nancy Cartwright argues that the sciences, including physics, require a notion of causal capacities (Cartwright 1989; 1999). Much more common, however, at least among philosophers of physics, appear to be positions that agree with Kirchhoff's or Mach's skepticism and maintain that metaphysically rich notions of causation can have no legitimate place in a mathematized empirical science such as physics. At most a "thin," broadly Humean notion of causation may be compatible with physics, without however playing any useful role within that science.

A second philosophical project aims to offer a conceptual analysis, broadly construed, of claims of the form "A causes B." The core criterion of success within the context of this project is that an account of causation be able to reproduce commonsense causal claims – that is, that it be able to match our intuitions regarding what is assumed to be our "folk notion" of causation. David Lewis and his followers are engaged in this type of project, for which the central data are commonsense claims such as "Suzy's throwing the rock caused the bottle to break." Assessing the success of a given analysis involves examining how well the analysis handles cases of preemption, late preemption, trumping, or overdetermination – all well familiar from the literature on Lewis's counterfactual analysis of causation (see, e.g., Collins et al. 2004). My aim in what follows is not to offer a conceptual analysis of causal claims. Commonsense causal judgments will be relevant to my discussion only insofar as I will try to show that certain characteristic features that have been taken to be central to causal claims both in common sense and the special sciences are not incompatible with physics.

Very often those pursuing the second project proceed by almost completely disregarding putatively causal claims in the sciences and the conditions under which such claims are asserted. An exception is philosophers who believe that commonsense causal claims can be grounded in, or can be reduced to, what is taken to be fundamental physics. A crucial role in such reductive accounts is usually afforded to the thermodynamic asymmetry that the entropy of a closed system does not increase. According to a tradition going back to Reichenbach (1956), causal claims – in particular the asymmetry of the causal relation – and the thermodynamic asymmetry have a common origin described by statistical physics. Barry Loewer (2007; 2008; 2012a; 2012b) and David Albert (2000; 2012), for example, defend an account of how it is that we possess a time-asymmetric concept of causal influence or control, by arguing that our commonsense concept tracks certain non-causal features of the world that are central to the foundations of thermodynamics. Loewer quite explicitly situates his account within Lewis's tradition of offering a broadly counterfactual analysis of commonsense causal judgments, arguing that an appeal to statistical physics can solve a problem Lewis's own theory has in accounting for the causal asymmetry.[3]

Loewer argues that counterfactuals reducible to the foundations of statistical physics are important to us because they "track the statistical mechanical probability distribution [grounding the entropy asymmetry] in ways that are important for the consequences of our decisions" (Loewer 2007, 323). There is a sense in which I agree with Loewer. The usefulness of causal relations in physics is intimately connected to a temporal asymmetry of our universe that can be captured in probabilistic terms. However, this connection does not imply that causal notions are reducible to non-causal features of physical systems. In particular, I argue that Albert and Loewer's attempt at such a reduction is unsuccessful. What we can learn about the relation between causal and statistical properties in physical systems does not allow us to distinguish between reductive accounts, such as Albert's and Loewer's, and metaphysically "richer" accounts of causation in physics, such as that of Maudlin (who appeals to the very same probabilistic asymmetries in support of his own account).

A third kind of project, finally, is what Woodward in his 2012 Presidential Address to the North American Philosophy of Science Association calls a "functional project" (Woodward unpublished). The functional project asks what if any the use of a certain concept is within a certain context. If it is to be legitimate to invoke causal reasoning in a certain domain, then causal notions have to be able to prove their usefulness in explanations or predictions, or in making our way about in the world. Thus, instead of asking for the metaphysical underpinnings of causal notions, the functional project asks what role, if any, causal notions play as part of our epistemic toolkit and as part of the representational resources. The legitimacy of causal notions or causal thinking is evaluated with respect to whether they serve a useful function, and any account of causation has to be defended with reference to the functional role of causal concepts.

---

[3] See Frisch (2005a) for a discussion of this difficulty.

Woodward takes his interventionist account (Woodward 2003) to be an example of this kind of project, arguing that identifying relationships that are exploitable for manipulation or control is one of the central goals of causal thinking. Here Woodward is appealing to a thesis developed in an influential paper by Nancy Cartwright – the thesis that the distinction between causal relations and mere correlations is needed to be able to discern ineffective from ineffective strategies (Cartwright 1979). In order to know whether a certain course of action would be effective in bringing about a desired outcome, it is not enough to know various *correlations* between outcomes of the desired kind and other kinds of events whose occurrence we might take to be under our control. We also need to have *causal* knowledge.

Cartwright's example is that people who carry a life insurance policy from TIAA-CREF, a company whose customers are primarily educators, tend to live longer. Merely being told that a correlation between carrying the insurance and life expectancy exists does not yet allow us to determine whether purchasing the life insurance is an effective strategy for increasing one's life expectancy. Rather, we need to know the causal structure underlying the correlation: we need to know whether purchasing the insurance has an effect on longevity or if, more plausibly, the two factors have common causes, such as the high level of education of the insurance members or their access to good health care. In the latter case, purchasing the life insurance would not be an effective strategy for increasing longevity.

As Woodward argues, the distinction between mere correlation and causal relations can be fruitfully characterized in terms of possible interventions into a system. Roughly, if two variables are related as cause and effect, interventions into the cause variable provide a way of manipulating the value of the effect variable. By contrast, if two variables are correlated but not causally related, then interventions into one variable will not affect the value of the other variable.

The very long title of Woodward's PSA presidential address includes the promise to offer "a defense of the legitimacy of causal thinking by reference to the only standard that matters – usefulness." One advantage of a functional account is that any such defense is relative to a specific domain or context. A concept may serve a useful function in one domain but not in others. Thus, one can, as many philosophers do, believe in the usefulness of causal notions both in common sense and in how the special sciences represent the world and nevertheless deny that causal notions have a legitimate function in physics. Thus, causal skeptics point to a list of putative features of representations in the special sciences that show the usefulness of, and indeed the need for, causal notions there. These features include that representations in the special sciences are local, partial, temporally asymmetric, multiply realized and ceteris paribus (see, e.g., Loewer 2008). Woodward himself appeals to a similar list of features, which he takes to be absent from representations in physics, to argue for at least a tentative skepticism about the usefulness of causal notions in physics. Many others, however, have been much more forceful in their rejection of causal notions in physics. Directly denying any functional role for causal reasoning in physics, Huw Price and Brad Weslake, for example, argue that causal notions are "epistemically inaccessible and practically irrelevant" in physics (Price and Weslake 2009).

I argue that the causal skeptics' charge is unfounded and that a functional defense of causal reasoning can be given in physics as well. Yet my claim that causal notions have a function in physics carries with it no metaphysical implications and leaves the disagreement between Humeans and non-Humeans unresolved. Thus, I agree with Woodward, who maintains that a functional framework implies no metaphysical commitments beyond what he calls a "modest realism." Causal structures, I will argue, are an important part of the toolkit that we use to represent the world within the context of physical theorizing. That is, there is no good reason for not treating causal structures on par with other representational resources that we employ in physics, such as dynamical laws or other kind of constraints. This is compatible with a broad range of attitudes toward these structures, from a thoroughgoing metaphysical realism to more instrumentalist attitudes. Woodward's "modest realism" consists of nothing more "than the assumption that the difference between those relations that are merely correlational and those that are causal has its source 'out there' in the world (as philosophers like to say) and is not, say, somehow entirely the result of some activity of ours" (Woodward unpublished). This "realism" is compatible with the view that causal representations, like scientific representations in general, always are representations by *us* for a specific *purpose* in a certain *context*, without implying that how we successfully represent the world is entirely up to us. That is, it is compatible with causal representations being perspectival *in the very same way* in which scientific representation in general may be thought to be perspectival.

3. Properties of causal relations

What characterizes causal relations or structures? Here the functional project makes contact with the descriptive project: the relations and structures identified as playing a functional role are *causal* relations and structures precisely because they exhibit at least some of those features characteristic of our common sense or "folk" notion of causation. John Norton (2003, 17–18) presents the following list of what at least since David Hume's time have become a fairly standard list of properties associated by philosophers with our folk notion of causation.

First, causal relations satisfy "a principle of causality," which Norton takes to consist of the conjunction of the two principles "every effect has a cause" and "the same causes always bring about the same effects."[4] Second, the causal relation is generally assumed to be asymmetric. That is, if a is a cause of b, it is not the case that b also is a cause of a. Third, the causal relation is generally assumed to be temporally asymmetric in that effects do not temporally precede their causes. Fourth, causal relationships often are taken to satisfy certain so-called locality conditions. One type of locality condition stipulates that the relata of the causal relations must be spatiotemporally localized entities or quantities. Another type of condition demands that causes can act only locally or that causes cannot act at a distance. Fifth, it is common to make causal claims in contexts where one wishes to identify what Norton calls a "dominant cause." Causal claims are often made relative to a coarse-grained representation of the system of interest.

It seems to me beyond doubt that some such list of properties character-izes our commonsense notion of cause. Do similar causal properties play a role in physics? Now, as we saw earlier, Russell, following Mach, famously claimed that the word "cause" is simply no longer used in the advanced sciences. Yet, as has been pointed out repeatedly – for example, by Patrick Suppes (Suppes 1970) and by Christopher Hitchcock (Hitchcock 2007) – this claim is simply false. Contrary to what Russell maintained, "cause" and "causal" and related words are still frequently used in contemporary physics. A widely used textbook on classical electrodynamics even main-tains that a principle of causality is "the most sacred tenet in all of physics" (Griffiths 2004, 424). Yet, since the use of causal notions is hardly more regimented in physics than in everyday life, causal terms are used to refer to a variety of different properties in a variety of different contexts.

One of the few physicists who are careful explicitly to distinguish differ-ent aspects of the notion of cause in physics is Fritz Rohrlich. According to Rohrlich, there are three different meanings of causality in classical

---

4 David Hume, of course, distinguished these two principles carefully and also pointed out that the first principle is in danger of being trivially true if expressed as "every *effect* has a cause." A more careful formulation is "every *event* has a cause."

physics (all of which are closely related to aspects of our folk notion): "(a) predictability or Newtonian causality, (b) restriction of signal velocities to those not exceeding the velocity of light, and (c) the absence of 'advanced' effects of fields with finite propagation velocity" (Rohrlich 2007, 50). The third sense refers to an instance of the temporal asymmetry of the causal relation and is the requirement that disturbances of a field associated with a field source propagate into the future and not into the past. A survey of contexts in which causal talk is used in physics seems to confirm that these three dimensions of the notion of cause indeed play a particularly important role in physical theorizing: first, that causes determine their effects; second, that causes act locally; and third, that the causal relation is asymmetric and that this asymmetry is closely related to the temporal asymmetry.

The first two aspects, for example, are part of Erwin Schrödinger's "principle of causality," which is the requirement that "the exact situation at *any* point P at a given moment is unambiguously determined by the exact physical situation within a certain surrounding of P at any previous time, say $t - \tau$" (Schrödinger 1951, 28). Also, although Schrödinger does not stipulate that the causal relation must be asymmetric, his principle of causality only (asymmetrically) demands that the situation at P be determined by the state at an earlier time. Determinism together with a temporal asymmetry also constitute what Niels Bohr calls a "causal description," which rests on the "assumption that the knowledge of the state of a material subsystem at a given time permits the prediction of its state at any subsequent time" (Bohr 1948, 312). Similarly, Hermann Weyl also refers to a principle of determinism as the "the law of causality" (Weyl 1989, 40), while Max Planck characterizes the "law of causality" in terms very similar to Norton's principle of causality: "Everything that occurs, has one or more causes, which together necessarily lead to the event in question" (Planck 1937, 83, my translation).

In the first several decades after the quantum revolution, it seems to have been common to claim that quantum mechanics forces us to abandon causality. Bohr, for example, contrasted causal descriptions with the non-deterministic descriptions of the new quantum physics, and he took the latter to pose a threat to causality precisely because it is an indeterministic theory. It seems, however, that we have since learned to live with genuine indeterminism. Not only have there been multiple philosophical treatments of probabilistic causation in the wake of Patrick Suppes's work (Suppes 1970), but it also appears to be much less common among physicists today than it was perhaps in the first half of the twentieth century to refer to a condition of determinism as *the* "principle" or "law" of causality. Thus, Rohrlich says that predictability is the hallmark of a particular type of causality – "Newtonian" or "classical" causality – which allows other, indeterministic notions of causality to exist as well. Nevertheless, the arguments I consider here have generally been proposed in the context of deterministic micro laws, and I do the same in my discussion here.

Dynamical causal locality constraints considered in physics fall into two broad classes: first, there are constraints against "gappy" causation, which take the form of prohibitions against causes acting across spatial, temporal, or spatiotemporal gaps; and, second, there are constraints on the speed of causal propagation (see Frisch 2005a). The two kinds of constraint are logically independent. One the one hand, we can imagine action-at-a-distance theories that allow for propagation across gaps, but nevertheless impose a maximum speed of propagation. Wheeler and Feynman's infinite absorber theory of classical electrodynamics is an example of such a theory. On the other hand, there can be theories that impose no maximum speed on causal propagation but do not allow influences to propagate across gaps.

Relativistic field theories satisfy both kinds of constraint: the presence of the field ensures that causes do not act across gaps, while relativity theory posits a finite upper limit on the speed of causal propagation. In fact, relativistic theories satisfy two distinct constraints. First is the condition that there is a finite, invariant velocity – the velocity of light. This condition is often expressed as demanding that spacetime have a lightcone structure. Second is the condition that there is no propagation in matter faster than the speed of light. Both constraints and the spacetime structures satisfying them are usually characterized in causal terms in the literature. For example, two points in spacetime that can be connected by a signal traveling at most at the speed of light – that is, points that are either timelike or lightlike related to each other – are called "causally connectable." Also, curves in spacetime representing points moving at less than or equal to the speed of light are called "causal curves." In quantum field theories, relativistic constraints are implemented in the form of a condition called "micro-causality," which demands that the commutator between fields at spacelike-separated spacetime points vanishes. Micro-causality is meant to capture the intuitive condition that the value of the field at one spacetime point can make no difference to the value of the field at another point, if the spacetime points are spacelike separated – that is, the two spacetime points could not be connected by a light signal or by any object moving more slowly than the speed of light.

Locality constraints, especially in the form of relativistic constraints, provide one of the two main applications of causal terms in contemporary physics. What role does the use of causal language play here? Norton has argued that these constraints cannot plausibly be understood to be part of a general causal principle, since this would imply that

> any theory not complying with the causal principles of modern physics is causally deficient. The immediate consequence is that older theories, notably Newton's mechanics, were causally defective in not admitting a finite upper bound to speeds of propagation. And that has the odd consequence that we were mistaken for hundreds of years in extolling the causal perfections of Newtonian mechanics. (Norton 2007, 223)

We must be careful, however, to distinguish the general framework given by Newton's *laws of motion*, which survive in amended form even in relativistic theories and which, as Norton points out, are indeed often taken to be paradigmatically causal, from the particular *force law of gravitational attraction*. As Norton himself emphasizes, worries about the latter, which is an action-at-a-distance law and violates both kinds of locality constraint that I distinguished earlier, go back even to Newton himself.[5] Although it is true that physicists came to accept the law of gravitational attraction in light of its astounding empirical success, it may be that conceptual worries about the law were pushed into the background rather than being successfully resolved. Thus, rather than pointing to Newton's law of gravity as a counterexample to a general condition of causal locality, one might instead take the development of relativistic theories as a vindication of the "ancient" demand that causes act locally. Yet this does not undermine the intuition that Newton's second law, in either its classical or relativistic incarnation, is the paradigm of a causal law. That is, the putative "causal perfection" of Newtonian mechanics concerns Newton's second law and its interpretation as identifying forces as the cause of motion, but not the law of gravitational attraction, which as an action-at-a-distance law might be thought to be causally defective.

Norton further argues that even when a causality condition is taken to provide a universal constraint, it functions only as a label for what

---

[5] At one point Newton expresses his own reservations about his theory of gravitational attraction as follows:

> That one body may act upon another at a distance through a vacuum without the mediation of anything else [...] is to me so great an absurdity that, I believe, no man who has in philosophic matters a competent faculty of thinking could ever fall into it. (In a letter to Richard Bentley, dated February 25, 1693, reprinted in Cohen 1978, 302–3].)

ultimately are purely formal mathematical constraints. Calling such constraints "causal" does no additional work. For example, the content of the
condition of micro-causality is exhausted by the claim that the commutator
of spacelike-related field operators is zero. Causal notions are dispensable in
this case, since the causal condition does not imply a commitment to any
structure in addition to the one captured by the functional dependencies
expressing that condition.

Yet one may object that Norton is here invoking a "damned-if-you-do,
damned-if you-don't" argument.[6] Advocates of causal principles seem to
face the following dilemma: if a given causal principle is formulated only
vaguely and imprecisely, then this shows that the principle is not properly scientific, as Kirchhoff and Mach have argued; but if the principle
can be given a precise mathematical formulation, then this shows that the
causal notions are dispensable, since we can use the precise mathematical
formulation of the principle on its own without relying on its causal interpretation. But we need not accept the second horn of Norton's dilemma
and insist instead that the mathematical statement is an expression of
a causal principle. The mathematical statement gains its plausibility not
least because it expresses a causal constraint; conversely, the very fact that a
causal principle can be expressed mathematically supports its scientific legitimacy.

The above provides only a brief sketch of some of the issues arising for
causal notions as locality constraints in physics. However, causal locality
conditions are not the main focus of my investigation here. Although one
important usage of causal language in the context of relativistic theories
is to express the time-*symmetric* locality constraint that spacetime has a
lightcone structure and that there is no propagation outside the lightcone,
there is a second aspect to causal talk in relativity: causal notions are also
used to mark a time-*asymmetric* distinction between the future lightcone,
which is called "the causal future" of an event, and the past lightcone,
which is the "causal past" of an event.

More generally, a condition of local causality in physics is often introduced as the time-asymmetric condition that the fields at a point are fixed
by the fields in the causal past of the point. This is in accord with the fact
that the causal relation is asymmetric: if c is a cause of e, then it is not
the case that e is a cause of c. Both Bohr's and Schrödinger's principles
reflect this fact: According to Bohr's assumption, a causal description of
a deterministic system is one that characterizes the system's evolution in

---

[6] I owe this point to Jim Woodward.

terms of an initial-value problem. An "anti-causal" description, by contrast, would then be one that describes the evolution of a system in terms of final-value problem. Schrödinger's condition of local determination is likewise a principle of past-to-future determination.

This third characteristic of the causal relation – its asymmetry – is arguably the most central of the three dimensions and is the main focus of my investigation. The asymmetry is clearly an integral part of our intuitive idea that causes "bring about" or "produce" their effects, but it is also an integral part of less metaphysically "weighty" notions of cause. As the identification of the causal future with the future lightcone in relativity theory attests, the causal asymmetry is intimately related to a temporal asymmetry, even though the precise nature of the relation is a somewhat delicate issue. On some accounts of causation, such as Humean regularity accounts, it is a conceptual truth that effects do not precede their causes. In the *Enquiry on Human Understanding*, Hume defines a cause as "an object, followed by another, and where all objects similar to the first are followed by objects similar to the second" (Hume 1975, 76). That is, for Hume a cause by definition precedes its effect. However, even those accounts that allow for the conceptual possibility of backward causation would presumably maintain that causation in our world (or at least in the spatiotemporal region of the universe accessible to us) is forward directed, and hence, causal constraints are often taken to imply time-asymmetric constraints (see Frisch 2013).

Norton, as we have seen, argues that the use of causal principles amounts to mere "labeling" and, hence, that we cannot draw any substantive conclusions from the use of causal language in physics. In addition to the "damned-if-you-do, damned-if-you-don't" argument criticized earlier, he offers the following argument. In general relativity, subsets of the solutions to Einstein's field equations are classified as causal in various senses. For example, solutions that permit closed timelike curves are classified as non-causal. They are non-causal because they violate a constraint that is usually imposed on asymmetric causal structures: the constraint that causal structures are acyclic. Yet, as Gödel was first to show, there are solutions to the field equations that contain worldlines for material particles that nowhere travel faster than the speed of light and yet are closed and return to their starting point. Thus, causal notions in this context are used to pick out proper subclasses of the theory's models. According to Norton, this suggests that causality conditions are nothing but devices for cataloging different solutions to the field equations. In particular, the causality conditions cannot be understood as additional factive constraints on physically possible solutions, since it is "routine to consider solutions to the Einstein equations that do not conform to them" (Norton 2007, 228).

An implicit premise in this argument, however, is the assumption identification of models of the field equations with the range of what, in the context of gravitational theories, is physically possible. Someone who wished to defend the view that a relativistic causality condition provides a factual constraint could deny that all models of Einstein's equations represent physically possible universes and could insist that causality conditions present additional potential constraints on what is physically possible. Norton's observation that physicists also consider models that violate the condition is no objection against this view. If a causality condition has the status of a hypothesis that is not yet sufficiently well confirmed, then investigating what the world would be like if the condition failed may be an important component of testing the condition. Furthermore, even if we took it to provide a well-confirmed constraint, exploring situations that violate the constraint – and hence are taken to be unphysical – can be a useful and fruitful exercise, because it might help us better understand the theory.

One might think that there could be no causal constraint in general relativity prohibiting closed causal curves – so-called causal loops – since there are strong arguments suggesting that the possibility of causal loops need not result in inconsistencies. However, any argument along these lines is in danger of confusing physical possibility with conceptual possibility. It may be the case that causal loops are *conceptually* possible yet *physically* impossible and that causal constraints are proposed as constraints on what is physically possible. A defender of the facticity of causal constraints, thus, will insist that we carefully distinguish among conceptual possibility, causal possibility, and what is possible according to a well-confirmed theory's basic equations.

4. Russell redux

Russell's essay "On the Notion of Cause" has had and continues to have an extraordinarily strong influence on philosophical discussions in the English-speaking world concerning the role of causation in physics. This is so despite the fact that much of Russell's criticism merely repeats arguments made earlier by Mach and others and despite the fact that Russell himself later abandoned his skepticism concerning causal notions. Because Russell's later views are much less well known, I briefly sketch them here (see also Eames 1989).

In *The Analysis of Matter*, Russell maintains, contrary to his earlier view, that "all science rests upon induction and causality" (Russell 1954 [1927]). Russell there rejects phenomenalism and argues for a causal theory of perception. Yet a partial skepticism about our knowledge of the external world remains. Since we have direct access only to our experiences – to what Russell calls "percepts" – Russell argues that all that we can know about the physical world are its structural properties:

> Whatever we infer from perceptions it is only structure that we can validly infer; and structure is what can be expressed by mathematical logic. (1954, 254)

> The only legitimate attitude about the physical world seems to be one of complete agnosticism as regards all but its mathematical properties. (1954, 270)

> If physics is concerned only with structure, it cannot, *per se*, warrant inferences to any but the structural properties of events. (1954, 390)

Consequently, causal relations ought to be thought of as structural as well:

> The aim of physics, consciously or unconsciously, has always been to discover what we may call the causal skeleton of the world. (1954, 391)

Russell's structuralism was attacked by the mathematician M. H. A. Newman for having the consequence of rendering the content of physics (almost) empty (see Demopolous and Friedman 1985). The problem is that we can define any arbitrary structure over any domain of objects, as long as the domain is large enough. Thus, the claim that "there is a relation R such that the structure of the external world with reference to R is W" (Newman 1928, p. 144, quoted in Demopolous and Friedman) is a claim only about the cardinality of the domain of objects, since "any collection of things can be organized so as to have the structure W, provided there are the right number of them" (*ibid.*). I will briefly return to this issue in the next chapter, but this is not our main concern here.

In 1927 Russell clearly no longer believed that the notion of cause had no useful role to play in philosophical accounts of physics. What, then, did he take the causal structure of the world to consist of? He proposes an account on which a relativistic spacetime and causal relations are interdefinable and causality "in its broadest sense" is understood "as embracing all laws which connect events at different times, or, to adapt our phraseology to modern needs, events the intervals between which are time-like" (Russell 1954). This suggests that what he might mean by "causal structure" is the structure exhibited by relativistic spacetimes, corresponding to Rohrlich's second meaning of causality in physics. However, Russell also presupposes that the causal relation is asymmetric. First, the relation between percepts and the physical events that caused them is asymmetric in a causal theory of perception: we can draw inferences from the structure of our perceptions to the structure of the physical world, since physical events cause our perception. Moreover, in *The Analysis of Mind*, Russell argues that such an asymmetric relation between percepts and their causes is not dependent on one of the two relata to be experiences and also exists between purely physical events:

> In order to eliminate the reference to our perceptions, which introduces an irrelevant psychological suggestion, I will take a different illustration, namely, stellar photography. A photographic plate exposed on a clear night reproduces the appearance of the portion of the sky concerned, with more or fewer stars according to the power of the telescope that is being used. Each separate star which is photographed produces its separate effect on the plate, just as it would upon ourselves if we were looking at the sky. If we assume, as science normally does, the continuity of physical processes, we are forced to conclude that, at the place where the plate is, and at all places between it and a star which it photographs, something is happening which is specially connected with that star. It must be something specially connected with that star, since that star produces its own special effect upon the plate. (Russell 1921, 99–100)

The star asymmetrically causes the image on the photographic plate. Indeed, Russell characterizes this asymmetry using a language that suggests a metaphysically rich notion of causal production:

> Thus in the case of a perception or photograph of a star, the active place is the place where the star is, while the passive place is the place where the percipient or photographic plate is. (Russell 1921)

Yet he insists in a footnote that not too much should be read into his use of the word "activity": "I use these as mere names; I do not want to introduce any notion of 'activity'" (Russell 1912, 130). Russell suggests that our grouping together different "happenings" associated with different images of a star in one location as the location of the star – that is, our association of different images with the star as common source – follows from an assumption of continuity or locality. That, however – as we will see in Chapter 5, where I will return to a detailed discussion of Russell's example of stellar observations – is not enough. I will argue that (for reasons not mentioned by Russell) our inference from the image on the photographic plate to the existence of a star is indeed a paradigmatically causal inference, but a crucial role here is played by the asymmetry of the causal relation.

5. Things to come

My aim in this book is to show that, contrary to what appears to be the received wisdom among philosophers of physics, causal structures play a legitimate role in physics. I will show, negatively, that widely accepted skeptical arguments are unsuccessful and, positively, that there are good reasons for allowing causal relations to play a role in physics. A broader question in which I am interested is how we represent the world within the context of physical theories. One aspect of this question concerns the claim that physics plays a special and privileged role in science. Thus, many of the negative arguments I will discuss aim to draw a distinction between the special sciences on the one hand and physics on the other. I will argue that the contrasts the causal skeptics evoke do not exist. Representations in physics are rather more similar to representations in other sciences than the skeptical arguments allow. Scientific representation *in general* is partial, coarse-grained, and context-dependent. These deeply pragmatic features of scientific representation are present in physics no less than in other sciences.

A second aspect of the question of how we represent the world within the context of physical theories concerns the role of laws. In one view, our representational resources in physics are all but exhausted by our theories' laws. What a physical theory presents us with, according to that view, is a set of models or possible worlds defined by the theory's basic equations – the theory's laws. Paradigmatic dynamical theories, in this view, define a well-posed initial-value problem and particular models of the world are defined by the conjunction of specific initial (or final) conditions and the dynamical laws. A theory is applied to a real-world system to make inferences from the state of the system at one time to its state at other times by determining the system's state on an initial-value surface and feeding that state into the theory's laws. I argue that this picture does not exhaust the representational resources of our theories and that without in addition positing asymmetric causal structures, many of the inferences we routinely make in physics would simply be impossible.

What do I mean by asymmetric causal structures? I will not offer a fully fleshed-out account of causation. However, the causal structures that play a role in physics can be represented (at least in a coarse-grained way) by directed acyclic graphs ("DAGs") that satisfy the causal Markov condition, which states that every variable in the DAG is probabilistically independent of its non-descendants conditional on its parents. To the extent that causal claims in physics can be represented in terms of a Bayes net formalism or structural equations, as developed by Spirtes et al. (2000) or Pearl (2000), this provides an immediate reply to one common anti-causal argument – the charge that causal claims are too vague to play a legitimate role in physics. In addition to the philosophers and physicists quoted earlier, this charge is even made by philosophers who otherwise are quite sympathetic to the importance of causal notions in our cognitive architecture. An example is Chris Hitchcock, who maintains that in physics "there are advanced stages in the study of certain phenomena when it becomes appropriate to eliminate causal talk in favor of mathematical relationships (or other more precise characterizations)" (Hitchcock 2007, 56). However, approaches such as Pearl's structural account of causation offer just such a mathematically precise characterization of causal relationships. Moreover, we will see that there is a mathematical machinery associated with a theory's dynamical equations, the so-called causal Green's functions, that can readily be integrated into a structural-equation approach to causation.

In Chapter 2 I will sketch a general pragmatic account of scientific representation, drawing on Bas van Fraassen's account, which will set the stage for some of the arguments in subsequent chapters and is meant to serve as an antidote against an overly metaphysical reading of my arguments defending causal reasoning in physics. Chapter 3 will examine two clusters of arguments advanced by causal skeptics that point to what are taken to be central features of causal representations that are compatible with the special sciences but not with how physics represents the world: causal representations are (i) coarse-grained and (ii) centrally involve a distinction between causes and background conditions. Neither of these two arguments can achieve their goal and show that representations in physics cannot involve causal relations.

On a Bayes-net or structural equations approach to causation, conceptual connections between the notions of causation and manipulation or intervention are especially perspicuous, as Pearl stresses in his work. However, there is a cluster of arguments intended to show that an interventionist notion of causation is incompatible with representations in physics. I will critically examine these in Chapter 4. Chapter 5 addresses what arguably is the most prominent strategy to argue against the legitimacy of causal notions in physics. This strategy appeals to the asymmetry of the causal relation and contrasts this with the putative fact that the dynamical equations of our fundamental theories are time-symmetric. I argue that several different ways of pursuing this strategy fail. I also argue, positively, that a time-symmetric notion of causation plays an important role in inferences in physics. Severe limitations on the data we have available to draw inferences from one time to another result in an underdetermination problem that can be solved only with the help of causal assumptions.

Chapters 6 and 7 examine two examples of time-asymmetric causal reasoning in physics: linear response theory and classical radiation theory. The source of the asymmetry of radiation has been controversially debated at least since a well-known debate between Albert Einstein and Walter Ritz. Because this debate is widely cited in the contemporary literature, where Einstein's and Ritz's positions are generally misrepresented, my approach in Chapter 7 will be partly historical, with the aim of correcting what have been misunderstandings of Einstein's and Ritz's views. In Chapter 8 I contrast my account with the reductive accounts of causation proposed by David Albert and Barry Loewer and by Huw Price. Chapter 9 summarizes my conclusions.

Anti-causal arguments often aim to establish that there is no role for causal notions in what is said to be "fundamental" physics. I end this chapter with a brief comment on what is meant by "fundamental" in this context. The arguments I want to consider here are meant to invoke features that apply quite broadly to any physical theory that is "on the books." That is, the arguments are meant to apply to any theory that is taught to physics students today and is applied in representations of physical systems. The arguments I will consider here do not appeal to any of the apparent peculiarities of quantum theories and are intended to include classical mechanics and classical electrodynamics within their scope. Because the arguments I will consider here are meant to apply to classical, that is, non-quantum, theories as well and are often discussed in the context of classical physics, I will also focus on these theories in my discussion here and will steer clear of arguments involving the interpretation of quantum theories. Because the theories in question are not truly fundamental, I will use the term "established theories" instead of "fundamental theories."

CHAPTER 2

Users, structures, and representation

1. Introduction

My aim in this chapter is to defend a pragmatic and structuralist theory of scientific representation and to explore certain consequences of such an account. The central tenets of the view I will explore are the claims that representation is an essentially pragmatic notion, and that representation, at least in physics, is structural representation. In recent years this view has been eloquently defended by Bas van Fraassen (van Fraassen 2008), and in what follows I will take his account as my main point of departure.

In the next section I briefly introduce and motivate a pragmatic account of representation according to which the link between a representation and its target is forged by a context of use. In Section 3 I introduce the second, structuralist plank of van Fraassen's account and discuss two criticisms of a structuralist account of scientific representation. First, I defend structuralism against the view that the content of a representation cannot be purely structural but needs to be concretely fitted out. Second, I briefly discuss Putnam's model-theoretic argument and the triviality objection against structural representation to which it gives rise. The main lesson I want to draw from Putnam's argument is the need to introduce a further pragmatic element into our account of representation: not only are target and content of a representation determined by a context of use, but the target itself is structured by the user in a given context for certain representational purposes. In Section 4 I turn to an objection to the pragmatist plank of the account that threatens to constitute a reductio of the view: if both target and content depend on the user of a representation, who provides a selective structured depiction of a phenomenon relevant to a particular context of use, then one might worry that the domain of scientific theories is radically restricted to only those phenomena for which explicit models have been constructed. Although this strong objection can be met, a weaker version of the argument goes through: putatively fundamental micro theories do not represent higher-level macro phenomena, or so I shall argue. There can be no fundamental theory of everything. My strategy in Section 4 is to begin by offering an argument for anti-foundationalism. I then show that the argument does not generalize to all phenomena for which no actual models have been constructed and, thus, that the anti-foundationalist argument cannot be broadened in scope to a reductio of a strongly pragmatic account of scientific representation.

Whereas some defenders of a pragmatic view of representation emphatically embrace anti-foundationalism (see, e.g., Teller 2001), others, such as van Fraassen, appear reluctant to endorse it. Structuralists in philosophy of science also are not all anti-foundationalists. But if what I argue here is correct, then structuralism (together with the denial of a strong essentialism) implies a pragmatic account of representation – an account of representation that also strongly recommends itself for independent reasons – and pragmatism implies anti-foundationalism.

The role of the present chapter, which does not explicitly make reference to causal structures, in my overall argument is the following. First, my defense of a structural view of scientific representation *in general* provides additional support for a structural theory of causation, such as the one proposed by Judea Pearl, which, as I will argue, is best suited to account for the role played by causal relations in physics. Second, I will argue in the following chapters that some of the putatively distinctive features of causal representations, which some skeptics of causation take to be characteristic of representations in the special sciences, are in fact features of scientific representation in general. Finally, I hope that my defense of a pragmatic account of scientific representation here will guard against an overly metaphysical reading of my arguments defending causal reasoning in physics in later chapters. One of my overarching theses in this book is that causal relations, just like theoretical laws, are part of the representational toolkit in physics. But this does not imply a commitment to a rich causal metaphysics.

2.   "No representation without representer"

What is it for a scientific representation to represent a phenomenon? Bas van Fraassen argues, to my mind convincingly, that representation is an essentially pragmatic, user-dependent notion that we cannot "define" or "reduce . . . to something else" (2008, 7). To call a thing a representation is to say something about its use. Thus, the "*Hauptsatz*" of van Fraassen's account of representation is that "*There is no representation except in the sense that some things are used, made, or taken, to represent things as thus and so*" (2008, 23, italics in original). This implies that there can be no "natural representations" – no naturally produced objects or phenomena that represent other phenomena without being used by someone to represent. Independent of its actually being used as a representation, a picture has no representational content: "*To call an object a picture at all is to relate it to its use*" (2008, 25, italics in original).[1]

I do not here want to defend the view that we cannot give a more substantive account of representation than what is expressed in van Fraassen's *Hauptsatz*, but I do wish to claim that the *Hauptsatz* will have to be part of any satisfactory account of scientific representation. Thus, one cannot define representation independently of use exclusively in terms of likeness or resemblance, for several obvious reasons. As has often been pointed out, at least since Nelson Goodman's seminal work (Goodman 1976), resemblance is a symmetric relation, but representation is not. Moreover, whereas perfect resemblance would be much too strong a requirement for representation, partial resemblance is much too weak: arguably for every two objects there will be some respect in which the two objects resemble each other. Hence, partial resemblance cannot be sufficient for representation, for otherwise we would be forced to the conclusion that everything represents everything else. But partial resemblance as necessary condition, without additional constraints, would be an empty requirement. Thus, according to a pragmatic account, representation is best thought of not as a two-place relation between a representation and its target but rather as (at least) a three-place relation, which includes a place for the user of the representation (and perhaps additional places for context, aim, or purpose): *a* is a representation of *b*, exactly if a user *u* uses *a* to represent *b*.[2]

It might seem that by insisting on the pragmatic character of representation, I am thereby disagreeing with recent structuralist accounts of representation, such as the partial structure account proposed by Steven French and Otávio Bueno (see, e.g., Bueno and French 2011), or the homomorphism account defended by Andreas Bartels (Bartels 2005; 2006).[3] Pragmatists

---

[1] A similar non-reductive account of representation is defended in Suárez (2004) and Suárez (2010). See also Suárez's and van Fraassen's contribution to Ladyman et al. (2011).

[2] Ronald Giere has proposed to understand scientific representation in terms of the following four-place relation: "*S* uses *X* to represent *W* for purposes *P*" (Giere 2006, 60). I take Giere's proposal to be implied by my perhaps slightly broader suggestion: purposes can be understood to be given by contexts.

[3] A homomorphism is a structure-preserving map. If the homomorphism is one-to-one, then it is an isomorphism.

and structuralists seem to be at odds concerning the nature of scientific representation. But French and Bueno also acknowledge an important role for pragmatic and context-dependent considerations. They stress that "partial isomorphism [that is, the structural relationship that they see at the core of successful representation] is not sufficient and that other factors must be appealed to" (2011, 29). Indeed, they maintain that just as a certain structural relationship is not enough to fix a representation's target, one also cannot rigidly build a particular intention into the representational mechanism that would permanently fix the representation's target. Rather we must allow for the flexibility of "pragmatic or broadly contextual factors to play a role in selecting which [representational] relationships to focus on" (Bueno and French 2011, 31).

Initial appearances to the contrary, Bartels also agrees with the claim that structural relationships are not sufficient for representation. Bartels argues that the common symmetry-objection to structural accounts of representation is unsuccessful and can be disarmed by taking the appropriate structural relationship between a representation and its target to be a non-symmetric homomorphism instead of a symmetric isomorphism. Bartels then proposes a homomorphism condition either as a sufficient condition (Bartels 2005) or as a necessary condition (Bartels 2006) for representation. Although this might be taken to suggest a purely structural account of representation, he ultimately distinguishes the notion of *potential representation* from that of *actual representation* and maintains that only the former notion satisfies the homomorphism condition. Whether a potential representation is also an actual representation is determined by pragmatic and context-dependent factors.

The difference, then, between a formal, structural account, such as French and Bueno's or Bartels's accounts, and van Fraassen's account strikes me as one of emphasis: whereas French and Bueno stress the structural relationships that have to exist between a *successful* scientific representation and its target, van Fraassen focuses on the ineliminable pragmatic aspects of the representation relation and the insufficiency of *purely* structural relations in establishing a representation relation between a structure and its target. And like French and Bueno or Bartels, van Fraassen takes representation in physics to be primarily structural representation. That is, all parties to the debate agree that the *content* of representations, at least in the physical sciences, is purely structural, but that what the structural content and the target of a representational structural are is fixed by the context of use. Representation is purely structural, since the models or representations employed, at least in the physical sciences, are mathematical structures and the only relevant resemblance between mathematical structures and physical systems is structural resemblance. And the representational content of a model and its target are fixed by a context of use, since similarities between a representation and its object are not sufficient to fix target and content. To the extent that resemblance plays a role in representation, it does so as a function of the representation's use. For example, in certain contexts we identify a representation's target with the help of selective resemblances between representation and target. Yet which aspects are important in assessing the likeness between representation and target is given by the context in which the representation is used.

Thus, there is no tension between a structuralism about representational content and a pragmatic theory of representation. Indeed, as we will see in the next section, far from being in tension with each other, structuralism provides additional support for a pragmatic account of representation. There is, however, a conflict between a pragmatic account and the widely held view that a physical theory's representational content – what the theory says about the world – is given simply by a statement of the theory's basic equations, which are taken to define the theory's representational structures. That view, which Nancy Cartwright has derisively dubbed "the vending machine view" of scientific theories (Cartwright 1999) and which van Fraassen himself appears to have once held (see, e.g., van Fraassen 1980),[4] is inadequate precisely because it ignores the central role played by users of representations in determining both a representation's target and its content, as I will argue later.

Once, however, we have abandoned the vending machine view and accept that the models we use to represent the world with the help of laws or theoretical principles are not simply given to us as soon as we have been given the laws or principles, then this opens up a space for other modeling assumptions, including causal assumptions, to play a role in constructing representational structures.

3.   Representational structuralism

3.1 *Do structural models need to be concretely fitted out?* Physical theories provide us with mathematical representations of phenomena – that is, with abstract structures introduced with the help of mathematical equations. Successful theories, it is often believed, provide us with models that in

---

[4] See, however, van Fraassen's discussion of his earlier view in Ladyman et al. (2011).

some sense and to some degree resemble the phenomena they represent. What kind of resemblance can exist between mathematical structures and concrete goings-on in the world? The most plausible answer to this question, as we have seen, and the answer that van Fraassen endorses, is that successful theories provide us with models that are *structurally* similar to the phenomena they represent. The view has a long tradition in the philosophy of science, dating back at least to Heinrich Hertz's and Ludwig Boltzmann's *Bildtheorie*, and is expressed, for example, in Hertz's famous dictum: "We form for ourselves inner apparent images or symbols of external objects, and we do this in such a manner that the necessary consequents of the images in thought are always the images of the necessary consequents in nature of the objects pictured" (Hertz 1910, 1). According to Hertz's view, successful representation requires only that the relations among our images or symbols of external objects be structurally similar to the relations among the objects but not also that our images resemble the objects.

There is a line of argument, however, that suggests that there can be no purely structural representation in science – that is, that there can be no structural representation without non-structural features also playing role. Model systems even in physics, according to the argument, have to be thought of as imagined or hypothetical concretely fitted-out physical systems: as Roman Frigg maintains, they "do not exist spatio-temporally but are nevertheless not purely mathematical or structural in that they would be physical things if they were real" (2010, 253). Here I am adopting Cartwright's expression of a concretely fitted-out model: a model is concretely fitted out when a concrete description is added to the model's structural, mathematical skeleton. Frigg argues that in order to make sense of a structural resemblance between a model and its target, one has to assume that the target also exemplifies a certain structure. But, Frigg argues, "this cannot be had without bringing non-structural features into play" (2010, 254). Putting it in Hertz's terms, our images or symbols of objects cannot be purely structural.

Frigg's argument against a purely structuralist account of scientific representation proceeds in two steps. First, he argues that, since structures are abstract, structural claims about an actual physical system cannot be true unless some non-structural claims are true of the system as well. Second, he points out that the "descriptions we use to ground structural claims are almost never in fact true descriptions of the intended target system" (*ibid.*), from which he concludes that "the descriptions that ground structural claims (almost always) fail to be descriptions of the intended target system. Instead, they describe a hypothetical system distinct from the target system" (*ibid.*). Thus, theoretical models cannot merely be mathematical structures but are concrete, albeit merely imagined or hypothetical physical systems.

Frigg's first step begins by noting a point also made by van Fraassen and to which we will return later: structural resemblance is possible only between two structures, and hence the target of a representation also has to be taken by the user of the representation to be structured in a certain way: "in order to make sense of the notion that there is a morphism between a model system and its target we have to assume that the target exemplifies a particular structure" (Frigg 2010, 254). As van Fraassen puts it, appealing to a notion introduced by Patrick Suppes, structural resemblance is between a theoretical model and a *data model*, which is a structural depiction of the target system. Frigg then argues that a structural claim about the target system cannot be true unless some more concrete claim is true as well. For example, that three iron rods exhibit a certain abstract ordering relation may be true in virtue of the rods having different lengths. While there are many different concrete descriptions in virtue of which a structural claim about the target system is true, there always is *some* non-structural claim that is true if the structural claim is true.

So far so good. Another way of making what amounts to the same point is to insist that data models are abstractions from a more full description of the phenomenon or system at issue. Grounding the data model there needs to be what Cartwright calls a "prepared description" of the phenomenon in question (see also Bogen and Woodward 1988). The observation that our theories' contact with the world is not ultimately grounded in data models but in an underlying concretely fitted-out description of a phenomenon appears to be correct.

But what the current argument ultimately needs to show is not that *a structured depiction of the target system or data model*, against which the success of a theoretical model is judged, is always accompanied by a more concrete description, but that *theoretical models* used to represent the target system themselves are hypothetical concrete physical systems. And the claim that any structural depiction of the phenomena needs to be embedded into a concrete description of the phenomena does not imply that theoretical models, too, need to be concretely fitted out. The crucial step in the argument, thus, is the second step, which begins with the observation that the descriptions accompanying a model are almost never true descriptions of a phenomenon and concludes that this fact makes the introduction of hypothetical systems necessary.

That theoretical representations in general partially misrepresent their intended targets is by now widely accepted (see, e.g., Cartwright 1983; Teller 2001 for arguments for this view). But why does this imply that theoretical models cannot be purely structural? The argument appears to be this. Since theoretical models involve idealizations, the claim that a given physical system has a certain structure is not strictly speaking true: taken literally, the claim "System W has structure S" is not true of any real-world system. Thus, taken literally, the structural claim must describe a hypothetical system. But, from the considerations above, it follows that there can be no system for which a structural claim is true without a concrete, non-structural description being true as well. Thus, concretely fitted-out hypothetical systems are an integral part of our theoretical machinery.

I have two worries about this argument. First, it is not obvious that it follows from the fact that no structural claim is true of *a real system* unless some more concrete claim is true as well that the same holds for *hypothetical systems*. Why can't the imagined system used to represent the real system not be purely structural? Any real solid object, for example, has to be colored. But it is not obvious that it follows from this that when I represent a die as a perfect cube, the representation has to be taken to be concretely fitted out as having some specific color. Second, it is not clear why the fact that a description is false or idealized implies that it "fails to be descriptions of the target system." More plausibly, it seems to me, one might hold that even an idealized description of a system is a description *of that system* – it just is a description that we do not take to be completely true but that nevertheless plays a useful role in our understanding of the system. Consider as an analogy a caricature that depicts Barack Obama as having huge ears. It does not follow from the fact that the caricature misrepresents Obama that there is some hypothetical person whom the caricature represents completely truthfully. Similarly, it does not follow from the fact, say, that a prepared description represents a certain wooden beam in an idealized manner as perfectly rigid that the description describes a hypothetical rigid object. Rather, the description represents the *actual* wooden beam *as* perfectly rigid. That is, the target of the description is the actual wooden beam, even though what the description says about the beam is strictly speaking false.

One might respond that a proposal along these lines leaves the notion of idealized description unanalyzed and that any plausible attempt to analyze this notion would need to introduce the notion of a hypothetical system after all. But one way to cash out the idea that a description is idealized is by reference to some other description or representation of that very system. Thus, we might say that a description D of a system S is idealized relative to some other description D^* of S just in case D ignores certain features attributed to S by D^* or simplifies certain features attributed by D^*. Both descriptions literally describe the target system, but D is idealized relative to D^*.

Another way to express what I have argued is as follows. According to the version of the semantic view of theories advocated by Ronald Giere (1988), we need to distinguish the models constructed with the help of a theory from a theoretical hypothesis, which posits a certain relationship between a model and the real-world system it is intended to represent. If we take into account that models involve idealizations and abstractions, then a theoretical hypothesis will state that a real-world system is similar (in some specific sense) to the model in certain respects. That is, what the hypothesis *literally* says is that there exists an approximate similarity between the model and the real-world system. The structuralists' thesis is that the similarity at issue is purely structural. It is not clear why, simply because the similarity between the model and the world is only approximate, we also need to posit a hypothetical concrete system that the model represents completely accurately. Moreover, it is not clear why it should follow from the fact that real-world systems will satisfy concrete descriptions in addition to structural claims, that the structural model also has to be taken to be concretely fitted out by a hypothetical concrete model system.

Frigg suggests another argument in support of the claim that model systems cannot be purely structural, appealing to the fact that "scientists often talk about model systems as if they were physical things" (2010, 253). This is surely right, but it is unclear what lesson we should draw from this observation. One option might be to argue that this merely points to a surface feature of scientific practice and ought not to be understood literally. Another possibility is that what scientists understand by a model system or a theoretical representation differs widely from discipline to discipline, from context to context and even from scientist to scientist. In some contexts, especially in the more fundamental theories of physics, theoretical models might consist of purely structural, mathematical representations of the phenomena; in other contexts the models might be concrete yet imagined natural systems.

Nevertheless, a rapprochement between someone who insist on the importance of concretely fitted-out descriptions and the structuralist seems possible. One can agree that structural models in physics are very often or even in general concretely fitted-out. The concrete model system might serve heuristic or didactic purposes or even might be taken to play an important role in our understanding of the phenomenon the model represents. Even so one can insist that the *representational content* to which we are committed as resembling the phenomena we are modeling consists of purely structural models. Again we can express this point in terms of Hertz's dictum. Representing objects not merely in terms of abstract symbols but in terms of concretely fitted-out "images" may play an important role in physics, even though our representational commitments and the conditions of adequacy for any representation are purely structural.

Consider the planetary model of the atom, which may be best thought of as involving two kinds of models – a purely mathematical model that is taken to structurally resemble actual atoms, and a hypothetical physical system that is proposed as a useful concrete analogy of the system modeled. The way in which atoms are proposed to be like planetary systems is exhausted, it seems, by the structural resemblance postulated in the mathematical equations governing the Bohr atom. That is, one can grant that a concrete fitting out of structural representation plays an important role in physics, while nevertheless maintaining that the representational commitments of physicists ought to be understood as concerning the structural part of a model alone.

Peter Godfrey-Smith has argued that scientific modeling involves a "gradient of abstraction" (Godfrey-Smith 2009, 104). On the one end of the spectrum we find approximate or idealized descriptions *of a concrete target system*, without the introduction of fictional entities representing the target. On the other end of the spectrum are the mathematical models of theoretical physics, which sometimes are investigated as purely mathematical structures in their own right, to some extent even independently of possible empirical applications. Somewhere in the middle are fictional concreta – hypothetical physical systems – which provide concrete analogies to systems represented structurally.

The overall picture that emerges, as far as representation in physics is concerned, is not unlike the one Frigg depicts in Figure 1 (2010, 266), according to which a concrete hypothetical model system, which is assumed to exhibit a certain mathematical model structure, represents the target system. In Frigg's diagram, which is reproduced here as Figure 2.1, the representation relation between the model system and its target is called "t-representation" for "target representation." The model system itself is represented by a linguistic description, which Frigg calls, following Kendall Walton, a "prop" for the model system (and, hence, calls the representation relation between model and linguistic description "p-representation"). A

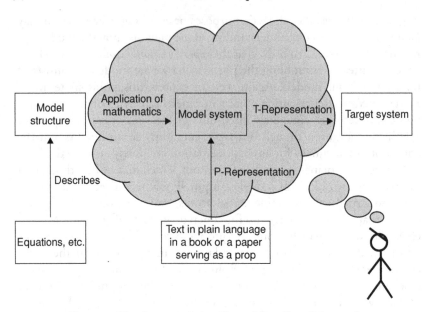

Figure 2.1 The elements of scientific modeling (from Frigg 2010)

structuralist could adopt this picture with one crucial change: the concrete hypothetical system and the model structure ought to trade places. For the structuralist, the model structure (represented by mathematical equations or other formal devices) occupies the central role in the framework and represents the target system. Concrete hypothetical model systems, by contrast, serve important heuristic or didactic purposes, but (at least in the heavily mathematized sciences) are not part of a scientist's explicit representational commitments (see Figure 2.2).

We can apply this framework to causal structures. The causal structuralist's claim is that time-asymmetric, acyclic relations, such as the ones at the heart of Pearl's structural theory of causation or of Spirtes, Glymour, and Scheines's causal Bayes net approach, play a representational role in science, including physics. These causal structures may be given a fuller concrete description, but the representational commitment is only to the causal structures and, in particular, entails no commitment to any particular causal metaphysics.

3.2. *The triviality objection.* Thus, the fact that scientific representation also involves concretely fitted-out prepared descriptions of a target system does not provide an argument against the claim that representation in physics is purely structural representation. There is, however, a second,

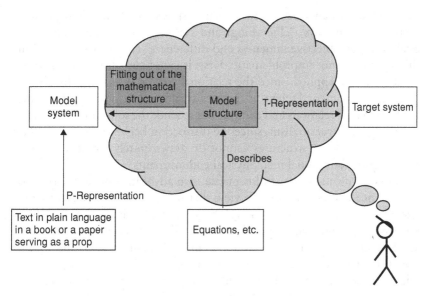

Figure 2.2 A structuralist account of scientific modeling

well-known problem for the view that scientific representation is purely structural representation: structural resemblance, it seems, is much too easily to be had. This point is expressed, for example, in Putnam's model-theoretic argument (Putnam 1978), which argues the following: As long as the physical system that we want to model is composed of – or is taken to be composed of – a sufficiently large number of parts, there will always be a mapping from the variables of the model onto parts of the system such that the system and the model exhibit similar structures. If there is no additional constraint that allows us to distinguish permissible from impermissible mappings, then the claim that there exists an appropriate structural resemblance between a model and some physical system turns out to be nothing more than a claim about the number of elements of the model and the system.

David Lewis replied to Putnam's argument by arguing that there is an additional constraint on the mappings given by preferred or natural divisions and relations among objects in the world (Lewis 1984). A representation is successful, according to Lewis's proposal, not merely if the physical system represented can be structured in some way that is isomorphic to the representation, but only if the representation is approximately isomorphic to the structure given by the natural relations among the entities. Thus,

Lewis advocates an anti-nominalism in response to Putnam's argument: there are, Lewis says, "elite things and classes," whose "boundaries are established by objective sameness and differences" (Lewis 1984, 227). But this solution is not unproblematic. First, it is not clear how generally the strategy can be applied, since the relational structures invoked by all but our most fundamental scientific theories are not good candidates for being the truly "elite" relations and things. And second, the solution moves away from a "pure" structuralism, since the distinction between natural or elite and gerrymandered structures cannot be drawn purely structurally and, thus, requires additional metaphysical commitments.[5]

Putnam's own solution to his puzzle is to advocate a deflationary view of reference combined with the view that our use of our representations fixes their meaning (see Frisch 1999). Yet rejecting Lewis's anti-nominalism appears to leave us with a problem, as van Fraassen points out: how can an abstract mathematical structure resemble something in nature that is not abstract? If we reject the idea that the world itself exhibits a preferred relational structure given by the elite things and relations, then what are the structures in nature that our scientific models are intended to resemble? Van Fraassen's answer to this problem points to an additional role of the user in representation. Theoretical models, he maintains, are intended to resemble *data models of the phenomena*, which themselves are constructed through the "selective relevant depiction of the phenomena by *the user of the theory* required for the possibility of representation of the phenomenon" (2008, 253). That is, the phenomena that our theoretical models are meant to represent are structured by us relative to our interests in a given context: "the phenomenon, what it is like taken by itself, does not determine which structures are data models for it – that depends on our selective attention to the phenomenon" (2008, 254).

The overall picture, then, that emerges is this. We represent a phenomenon by what van Fraassen calls a "data model" of the phenomenon and that provides a structured depiction of the phenomenon; a successful theory then provides us with theoretical models into which data models of the phenomena within the theory's domain can be approximately embedded, where it is our use of the theoretical models that singles out the intended embedding. Thus, the user enters the account of scientific representation at two places: first, in the depiction of a phenomenon as structured in a certain way; and second, in taking a model to represent

---

[5] See also van Fraassen's reply to Ladyman in Ladyman et al. (2011).

the phenomenon, depicted as thus structured, as having such and such features.

Finally, one might worry, as I argued above, that the first step in depicting a physical system is not yet a mathematical and precise data model, which often already involves a significant amount of theoretical analysis, but rather a still somewhat informal and concretely fitted-out prepared description. Taking this insight on board, we arrive at a picture that is slightly more complicated than the one suggested by van Fraassen himself and includes a prepared description, data models, and the models constructed with the help of a theory. In a given context we describe a system in a certain way that makes the system amenable to theoretical analysis. This prepared description guides our selective attention to certain aspects of the phenomenon in constructing a data model, against which a theory's models are tested.

To sum up, I have defended a pragmatic and structuralist account of representation against the objection that the content of representations in science can never be purely structural but always needs to be concretely fitted out. I have also suggested that answering a second objection – the triviality objection of Putnam's model-theoretic argument – introduces an additional pragmatic element into the account. According to the core thesis of a pragmatic account of representation, the link between a representation and its target is forged by a context of use. Meeting Putnam's challenges requires in addition that a representation's target be structured by a context of use.

4. Anti-foundationalism

In the previous section I considered two objections to the *structuralist* plank of a pragmatic structuralist account of representation. In this section I want to defend the account against an objection to the *pragmatic* plank. The account may seem to have the absurd consequence that the domain of a theory is radically restricted to those phenomena for which we have actually constructed a model or representation with the help of the theory. I want to argue here that this worry can be met. Yet, a restricted version of the putative objection is successful: a pragmatic structural account of representation is incompatible with scientific foundationalism, which is the view that physics aims to discover fundamental micro theories that have a universal domain of application and in principle possess models of all phenomena. The anti-foundationalist consequence, however, is no reason to reject the pragmatic account but rather a reason to embrace anti-foundationalism. I will approach these two issues in reverse order: First, I will argue for anti-foundationalism, defending the view against an objection suggested by van Fraassen's own discussion of a closely related question. Then I will discuss the worry that the anti-foundationalist argument generalizes and argue that this worry can be met without, however, providing solace to the foundationalist. Van Fraassen shows that there are close links between a pragmatism and a structuralism about representation in physics. I argue here that anti-foundationalism also has to be part of the "package."

There is a line of argument, developed for example by Teller (2001), for the conclusion that all scientific representations distort and partially misrepresent their intended targets. The argument points to the different and incompatible models physicists use to represent different aspects of the behavior of one and the same physical system. Teller's example is models of different aspects of the behavior of water. Continuum models can successfully represent the wave behavior of water, whereas particle models are used to represent diffusive behavior. The lesson of examples such as this is supposed to be that different types of model of one and the same system, constructed for different purposes, may make incompatible assumptions about a given system and that there is no single type of model that represents all aspects of a phenomenon adequately. Scientific foundationalists reply to this argument by maintaining that although the incompatible idealized and distorting macro models of water may be pragmatically useful, there exist perfectly accurate fundamental models, constructed with the help of our most fundamental theories, such as quantum mechanics, from which we could in principle derive the idealized higher-level models we use in practice. That is, foundationalism is committed to the view that our most fundamental micro theories have a universal domain and possess models of all phenomena, including higher-level macro phenomena, which we normally model using macro theories.

I want to distinguish two kinds of foundationalist views: strong foundationalism, which holds that macro theories are in principle dispensable and that we retain these theories merely for reasons of computational convenience; and weak foundationalism, which holds that even though fundamental physics in principle provides us with models of everything, macro theories nevertheless play an explanatorily indispensable role. Pragmatic structuralism is incompatible even with weak foundationalism. The reason is simply this: contrary to what the foundationalists assumes, we do not have fundamental models representing macroscopic phenomena. To actually construct a quantum-mechanical model of a macroscopic body of water, we would have to solve the Schrödinger equation for on the order of 10^{25} variables – something that is simply impossible to do in practice.

Foundationalists reply that even though it is impossible actually to solve the Schrödinger equation for macroscopic systems, the theory nevertheless provides us with models of arbitrary complexity. The equation defines a class of models, many of which we of course never construct explicitly. Indeed, any physical theory has many, many more models than the ones scientists have actually constructed and actually used. Quantum mechanics contains a model of the hydrogen atom, of Bose-Einstein condensates, but also, one might argue, of arbitrarily complex systems, including a glass of water. If a solution of the Schrödinger equation exists for complex initial conditions characterizing systems consisting of 10^{25} particles, then simply in virtue of being in possession of the *equation* we thereby also are given a *model* of such systems, even if we do not know how to explicitly construct this model.

But if a pragmatic account of scientific representation is correct, then the mere fact that an equation has solutions in addition to the ones actually constructed does not imply that whenever we possess a theory we thereby also are in possession of a large range of models of arbitrarily complex systems putatively governed by that theory. That quantum mechanics provides us with a model even of macroscopic bodies of water is supposed to follow from the claim that among the set of solutions to the Schrödinger equation are ones defining structures that are similar to bodies of water. Yet according to the pragmatic account, no structural relationship between a model and a phenomenon can on its own suffice to make the model a representation of that phenomenon (see van Fraassen 2008, 250). Rather something is a representation only if it is *used* to represent a thing. But since we do not even actually *have* the quantum-mechanical initial state of the system of water, let alone a solution of the Schrödinger equation for that system – since there is no way for us to pick out the appropriate solutions from the class of solutions defined by the Schrödinger equation – we cannot *use* the putatively existing solution to represent anything.

But is it not correct that the Schrödinger equation has many more *models* than the ones actually constructed by us? And if we allow that the Schrödinger equation has models for arbitrarily complex initial conditions, does it not follow that among the equation's models there will be some model with the right sort of initial conditions to represent the state of the body of water? This reply, however, trades on an ambiguity in the term "model." There are two quite different senses of model between which we have to distinguish carefully. On the one hand, there is the notion of *model as representation*, according to which a structure is a model of a thing if and only if it is used to represent that thing. That is, something *is a model of some object or system* by virtue of representing the object or system. And according to the pragmatic account of representation, nothing is a model in *this sense* without actually and as a matter of fact being used as a representation – the existence of a certain structural relation between the model and the target system is not enough for it to be a representation of the system. On the other hand, a structure is a *model of a set of sentences*, in the *logical* or *model-theoretic sense* of "model," if and only if it satisfies that set of sentences, that is, the set of sentences are true in that structure. A linguistic description of a theory serves to specify the theory's model-theoretic models in the sense in which a set of equations specifies the set of its own solutions. This second notion of model is not an intentional notion. All that is required for a structure to be a model in this sense is that a mapping from the structure to the set of sentences exists such that all the sentences in the set come out true; it need not be the case that there is a user who takes the set of sentences to be true in that particular structure.

If we accept van Fraassen's account of representation and that "there is no such thing as representation apart from or independent of our practice" (2008, 258), then it does not follow from the fact that a set of equations has solutions or models in the non-intentional, model-theoretic sense, which exist all along even without us using or being able to construct these solutions, that these solutions are also models in the first, representational and intentional sense. Recall van Fraassen's "*Hauptsatz*": "*There is no representation except in the sense that some things are used, made, or taken, to represent things as thus and so.*" Solutions to equations that we have not found or constructed cannot be used to represent anything, simply because we cannot use anything that we do not have or do not even know exists. Even if we assume that the Schrödinger equation has a solution for a system of 10^{25} variables of the kind that we might possibly use to represent a small macroscopic body of water if we were to possess this solution, the *mere existence* of the solution does not imply that there is a quantum-mechanical representational model of the waves in a body of water or of diffusion phenomena in water. Moreover, recall that the user enters on both "ends" of the representation relation, as it were: a model represents its target in certain respects and degrees only if a user takes it to represent the target in these respects and degrees; and the user gives the target a structured description appropriate for the theoretical representation in a given context. In the case of the hypothetical quantum-mechanical model of a cup of water, the use is missing on both ends. The idea, then, that our most fundamental micro theories provide us with representational models covering all physical phenomena is a foundationalist myth.

Surprisingly, van Fraassen himself does not draw this conclusion or at least is ambiguous in his views regarding this issue. He considers the question in what sense a theory or equation can be a theory of phenomena not actually encountered in practice – that is, of phenomena that we have not actually described and for which have not actually constructed a model with the help of the theory or equation. Van Fraassen would like to agree with what may be the common view and "would like to say that if the equation does have [an appropriate] solution – equivalently, if the theory has such a model – then that (equation, theory) does correctly represent that phenomenon" (2008, 250). Later in the book he says:

> The sense in which a theory offers or presents us with a family of models is just the sense in which a set of equations presents us with the set of its own solutions. In many cases, no solutions to a given equation are historically found or constructed for a very long time . . . though *mathematically speaking*, they exist all along. When the equations formulate a scientific theory, their solutions are the models of the theory. (van Fraassen 2008, 310, italics in original)

The latter passage occurs within the context of a discussion of Cartwright's view that there are models in science that have an existence that is in some sense independent of the theories with the help of which they are constructed. Thus, van Fraassen here appears to be guilty of not carefully differentiating between the two notions of model that I distinguished above: the solutions of the equations are *model-theoretic models* but not thereby automatically also *representational models*. Even though Cartwright is clearly interested in models as representational structures, van Fraassen invokes the model-theoretic notion of model in his response to her. Yet – putting the point without using the ambiguous term "model" – the fact that we can formally define a class of structures that satisfy a set of sentences says nothing about the representational use to which we might put those members of the class that we have explicitly constructed. And van Fraassen himself elsewhere in the book appears to stress this very point:

> There is no such thing as representation apart from or independent of our practice. So how can we say something like "this theory accurately represents that . . . phenomenon" although the relevant model was never constructed and the [phenomenon in question was] not encountered in human practice? The structural relationship between the model in question and the phenomenon . . . does not suffice to make the model a representation of the phenomena. (2008, 249)

And:

> Undoubtedly in many contexts something is called a model only if it is
> a representation, and the sense in which any solution of an equation is a
> model of the theory expressed by that equation certainly does not have that
> meaning. (2008, 250)

Thus, van Fraassen apparently wants to be committed to two ideas that
seem to be in tension with each other: on the one hand, the idea that "we
would like to say" that if a theory formally has an appropriate solution,
then it does represent the phenomenon in question even if the solution has
not been explicitly constructed; and on the other hand, the idea that there
is no representation independent of its being used as such and that "there
is nothing useful to be found in 2-place structure-phenomenon relations
alone when we try to understand representation" (2008, 258).

His resolution of the apparent tension is to understand the notion of the
empirical adequacy of a theory in terms of counterfactual representations:

> If the theory is offered, that amounts to the offer of a range of structures – the
> structures we call models of the theory – as candidates for the representation
> of the phenomena in its domain. If this range contains a candidate that
> would satisfy the structural constraint – if the phenomenon is embeddable
> in it . . . – then the theory is empirically adequate. (2008, 250)

That is, offering a theory amounts to offering a range of model-theoretic
models – of mathematical structures that we could use to represent phenomena. And a theory represents a particular phenomenon within its
domain adequately exactly if there is a structure among the range defined
by the theory such that *were we to use* this structures as representation
for that phenomenon, then the phenomenon *could* be embedded into the
model.

There is a certain irony in the fact that van Fraassen feels the need to
appeal to counterfactuals here, given his well-known view that counterfactuals are inherently and irreducibly context-dependent. What are the
truth-conditions of claims of the form "if we were to use a structure s to
represent phenomenon p, then p would be embeddable in s"? The problem is that it is not clear, independent of our actual use of a structure to
represent a phenomenon, how the structure would be used to represent
the phenomenon and what the appropriate embedding relation would be
were we to use the structure as representation. Thus, as van Fraassen himself emphasizes, we have to be careful about an "illegitimate slippage from
'there exists' to 'we have'" (van Fraassen 2008, 233). Although there may
exist solutions to the Schrödinger equation for systems of 10^{25} variables,

simply writing down the general form of the equation does not imply that we thereby *have* all of its solutions of arbitrary complexity.

Pace van Fraassen, pragmatic structuralism appears incompatible with scientific foundationalism. But one might worry that the argument goes far beyond an argument against foundationalism. Does a pragmatic account of representation imply that our theories can represent only those phenomena for which we have actually constructed models? This arguably would amount to a reductio of the account. In accepting Newtonian physics, say, we appear to be committed to the claim that the theory successfully applies to planetary systems yet to be discovered and systems of billiard balls never explicitly modeled. Even anti-foundationalists would want to insist, it seems, that any theory's domain extends well beyond the class of phenomena for which we have actually constructed models. How, then, can we combine this seemingly obvious point with the lessons of Putnam's argument and van Fraassen's Hauptsatz?

It seems to me that we need to distinguish carefully between two kinds of cases – extensions of the theory to phenomena of a kind that we have as a matter of fact already explicitly modeled and extensions that purport to go beyond this. The former extend a theory to hitherto not yet explicitly modeled phenomena or systems closely analogous to phenomena or systems already modeled. In the latter case, it seems to me, van Fraassen's strategy of counterfactual extension can be successful, even though it fails in the former case – the case needed by the foundationalist. Consider the example van Fraassen himself discusses in connection with this issue – that of a colony of bacteria located somewhere in Antarctica long before the first humans appeared on Earth.[6] Van Fraassen asks whether we can say that a theory of exponential growth adequately represents the growth rate of this colony, even though by hypothesis no model for this particular phenomenon was ever offered. His reply, as we have seen, is to appeal to a counterfactual account of empirical adequacy: the theory is adequate if among the solutions to its equations is one defining a structure that would satisfy the relevant constraints on adequacy if it were used to represent the colony's growth rate. The worry raised by Putnam's argument is whether this counterfactual has reasonably well-defined truth conditions.

I want to submit that that the answer is "yes" in the present case, since scientists actually and as a matter of fact use models of bacterial colonies

---

[6] In fact van Fraassen (2008, 25–6) discusses the putative worry as to how our models might be able "to represent something that has not yet entered our acquaintance." My worry here is, as it were, the mirror image of this concern: How can we represent anything with models that have not yet entered our acquaintance?

to represent their growth and arguably this practice sufficiently constrains how we would represent the Antarctic colony were we to do so. Scientists actually offer prepared descriptions of bacterial colonies and construct data models of the colonies' growth (consisting, for example, in plots of a colony's size at different times) that are appropriate for being matched with representation of the colonies' growth rate in terms of exponential growth models; scientists also actually use the latter models to represent bacterial colonies. Arguably, this practice sufficiently constrains what it would be to provide a data model of the Antarctic colony – that is, what it would be for us to selectively structure the phenomenon in a way that is relevant to exponential growth theory.

As van Fraassen emphasizes, however, the notion of relevance here is relative to a user and a specific context of use:

> *There is nothing in an abstract structure itself that can determine that it is the relevant data model, to be matched by the theory.* A particular data model is relevant because it was constructed on the basis of results gathered in a certain way, selected by specific criteria of relevance, on certain occasions, in a practical experimental or observational setting, designed for that purpose. (2008, 253, italics and emphases in original)

Because of this context dependence, there is a well-defined answer to what it would be to depict the Antarctic colony and embed its data model into a model of exponential growth only because the situation so closely resembles phenomena we have actually modeled and which can therefore provide appropriate criteria of relevance. Our actual modeling practices provide constraints on the truth conditions of what the structural constraint on our models would be in a counterfactual scenario.

The situation is dramatically different in the case of a putative quantum-mechanical micro model for the macro behavior of water – for the solution to the Schrödinger equation for 10^{25} variables. We do not actually have the relevant prepared descriptions for macroscopic bodies of water to be matched by a microscopic quantum-mechanical model, and we do not have actual examples of how a data model of a body of water might be embedded into a putative quantum-mechanical micro model. Thus, in this case we cannot draw on actual cases of modeling systems similar to the glass of water to constrain what it would be like to construct a quantum-mechanical data model for the glass of water. There is neither a well-defined answer as to what the relevant data model would look like, completely independently of any actual modeling practices, nor a well-defined answer as to what the appropriate embedding relation would be.

It is utterly unclear, then, what it would be for the range of structures defined by the Schrödinger equation to satisfy van Fraassen's condition of empirical adequacy – that is, for the range to "contain a candidate that would satisfy the structural constraint" that the phenomenon would be embeddable in it (2008, 250). The lesson of Putnam's argument is that there will be some mapping from the 10^{25} variables onto bits of the body of water such that the theory comes out true, no matter what its details are. And in this case there is no additional constraint – no practice of actually modeling the wave or diffusion behavior of macroscopic bodies of water microscopically – that can serve to single out an intended mapping.

In response one might try to appeal to our actual practices of modeling simple microscopic systems quantum mechanically as providing the relevant constraints, but what this reply ignores just how difficult it can be to construct an appropriate data model for a particular phenomenon. Consider as perhaps an extreme case data models constructed to test the standard model of particle physics at CERN. A lot of background theory and sophisticated statistical analysis goes into constructing a data model for classes of collision events each involving only a small number of elementary particles. It is unclear that we can have even the slightest idea of what it might mean to scale these models up to involve 10^{25} particles or more. We do now know how to extend, even in principle, "our decisions in attending to certain aspects [of the proton collisions at CERN], to represent them in certain ways and to a certain extent" (2008, 254) to provide us with a quantum field theoretic data model of a glass of water.

The pragmatist's *Hauptsatz* may suggest the following prima facie plausible account of what it is for a phenomenon to be within the domain of validity of a theory:

(D) A phenomenon P is in the domain of validity V of a theory T, exactly if

(i) T allows us to construct a structure M and
(ii) M has a substructure S, the intended representational structure in a certain context of use C, that is approximately isomorphic to P.

But the condition, as stated, falls prey to Putnam's argument: as long as the domains in question are large enough, there will always be an approximate isomorphism from S to P. Thus, we ought to replace (ii) with

(ii*) M has a substructure S, the intended representational structure in a certain context of use C, into which an (actual or counterfactual) data model D of P can be approximately isomorphically embedded in the manner that is intended in C.

I have argued that in extensions of a theory from phenomena that we have actually modeled (such as a bacterial culture in the lab) to what in a given context are relevantly similar phenomena (such as the bacterial culture in Antarctica), we know what data models of the not-actually-modeled phenomena would look like and what the appropriate embedding relation would be. By contrast, in the case of putative extensions of a micro theory to macro phenomena of a kind different from those explicitly modeled, we know neither what the appropriate data models nor what the appropriate embedding relations would be. That is, in the latter case the claim that a counterfactual data model could be embeddable into a substructure of one of the theory's models in what would be the intended way does not have well-defined truth conditions.

What phenomena count as relevantly similar is itself context-dependent, and the distinction between those phenomena that are within the proper domain of a theory and those that are not – that is, the distinction between those phenomena for which the counterfactual embedding relation is well-defined – is somewhat vague. The central reason, I have argued, for why van Fraassen's counterfactual criterion of empirical adequacy fails in the case of extensions of a putatively fundamental theory to every actual system and phenomenon is provided by the pragmatic response to Putnam's argument: the problem for the foundationalist is not merely that a context of use is needed to forge the link between a representation and its target, but that the target has to be structured in light of results "selected by specific criteria of relevance, on certain occasions, in a practical experimental or observational setting, designed for that purpose."

5. Conclusion

Structuralism, by way of answering Putnam's triviality objection, provides further support for a pragmatic account of representation – an account that recommends itself also for independent reasons. If a pragmatic structuralism about scientific representation is correct, then this has far-reaching implications for how we think about scientific theorizing. First, the view directly undermines "the vending machine view" of theories, according to which the representational content of a theory is given simply by stating a set of sentences or by defining a model-theoretic class of models, independent of a theory's users. Second, the view has radically anti-foundationalist implications. While our actual modeling practices sufficiently constrain what it would be to construct adequate models of other phenomena of a kind relevantly similar to the models we have actually constructed, our practices do not license an extension of our theories to every phenomenon, as posited by physical foundationalism. Finally, as we will see in more detail in the following two chapters, certain putatively characteristic features of causal representations in the special sciences, such as their context dependence and partiality, are in fact features of scientific representations in general.

CHAPTER 3

The human face of causation

1. Introduction

In the previous chapter I presented and partially defended a pragmatic account of scientific representations in general. In this chapter I will discuss a cluster of arguments that appeal to specific putatively pragmatic aspects of causal representations and are aimed at showing that it is precisely the fact that causal notions are essentially tied to our human perspective that makes their application in fundamental physics strained or problematic. As James Woodward puts it, causation has a "human face" and it is this human face of causal notions that results at the very least in a "failure of fit," "unimportance," or even, more strongly, a "disutility" (Woodward 2007) of causal notion in physics. The general argumentative strategy is quite popular and is pursued in various forms by Richard Healey (Healey 1983), Huw Price (Price 1997; Price and Weslake 2009), Norton (2003), Hartry Field (2003), Jim Woodward (Woodward 2007), and Christopher Hitchcock (Hitchcock 2007).

For several causal critics, the pragmatic aspects of causal notions not only provide a reason why causal notions can play no role in physics, but also can explain why causal thinking is important, despite the fact that, as they believe with Russell, causal notions play no role in physics. Hartry Field, for example, maintains that there is a prima facie puzzle created by the putative fact that the world is fundamentally a-causal, on the one hand, and by the insight, emphasized by Nancy Cartwright (see Chapter 1), that causal reasoning plays a central role in commonsense reasoning, on the other. Thus, for Field "the problem of reconciling Cartwright's points about the need of causation in a theory of effective strategy with Russell's points about the limited role of causation in physics . . . is probably the central problem in the metaphysics of causation" (Field 2003, 443). Field argues that one can do justice to Cartwright's insight by emphasizing aspects of causal reasoning that are essentially tied to our perspective as human agents interested in making our way about in the world. Once we recognize the human face of causation, we can do justice to the importance of causal representations without having to postulate that causal relations are woven into the fundamental fabric of the world.

Since many of the critics of causal notions in physics explicitly appeal to Russell's view in "On the Notion of Cause," I will call them the "neo-Russellians." But contrary to what Russell himself argued in this paper, many neo-Russellians do not advocate a wholesale rejection of causal notions. Rather, they grant that causal thinking plays an important role in commonsense thinking and the special sciences and account for the alleged absence of causal notions in physics by adopting a perspectival or pragmatic account of causation. Thus, extending Russell's famous analogy between causal notions and the monarchy (see Chapter 1), Price and Richard Corry argue for what they call *the republican option*: just as republicans take political authority to be vested in our rulers by us, causal republicans believe that "although the notion of causation is useful, perhaps indispensible, in our dealings with the world, it is a category provided neither by God nor by physics, but rather constructed by us" (Price and Corry 2007, 2).

The neo-Russellian arguments I will discuss in this chapter can be reconstructed as instances of the following general argument scheme:

(i) There is a certain (pragmatic/user-dependent/perspectival) feature x that is essential to causal representations.
(ii) Representations in physics do not have feature x.
(iii) Therefore, representations in physics cannot be causal representations.

I will examine several instances of this scheme and show that they all fail. The specific arguments I will consider in this chapter appeal to the fact that causal relations are coarse-grained and multiply realized (in Section 2) and that all causal claims are *ceteris paribus* claims relative to certain background conditions (in Section 3). In the next chapter I will examine arguments that appeal to another feature that pragmatic or perspectival accounts of causation stress: the close connections between the notion of causation and that of intervention, of the kind postulated, for example, by Woodward's account of causation. Perhaps the most widely cited anti-causal argument in this context appeals to the putative fact that the causal relation is time-asymmetric, whereas the dynamical laws of physics are time-symmetric. This argument will be the focus of Chapter 5.

One upshot of my discussion will be that the portrait neo-Russellians paint of causal representations resembles the face of scientific representation much more generally and that their arguments fail to establish a genuine contrast between reasoning in physics and reasoning in the special sciences. If causal representations have a human face, then so does scientific representation in general. Although the main thrust of my discussion will consist of challenges to different instantiations of premise (ii) of the general argument schema, I will also challenge certain instantiations of premise (i).

2. Coarse-graining

Paradigmatic commonsense causal generalizations are claims like "striking a match causes it to light." They are claims relating small numbers of events or event types, and the events at issue are relatively coarse-grained macro events. Implicit in commonsense causal claims is also a distinction between causes and background conditions. Intuitively, the striking of the match is a cause of the flame, while the presence of oxygen might seem not to be, even though both conditions are necessary for the match to light. Since the special sciences also concern relations between coarse-grained events and whatever regularities or mechanisms these sciences postulate are ceteris paribus and only hold relative to certain background conditions, the special sciences can apparently readily accommodate causal claims. This, according to Woodward and Field, contrasts sharply with the case of physics. The dynamical laws of a physical theory, stated in the form of differential equations, they maintain, require precisely specified initial states as input, defined over a global time-slice of the world or at least over an entire cross section of the past lightcone of the event of interest, allowing no distinction between background conditions and dynamically relevant factors: "In contrast to the incomplete relationships of limited invariance between coarse-grained factors that are characteristic of the upper level sciences, fundamental laws typically take the form of differential equations, deterministically relating quantities and their space and time derivatives at single spatiotemporal locations" (Woodward 2007, 83). What is more, the relations between different spatiotemporal locations are such that *"information about what happens at an earlier time can't suffice to determine the event unless it includes information about each point at that time that is within the past light cone,"* as Field emphasizes (Field 2003, 439, italics in original). I will examine both the coarse-graining and the background conditions claims separately, beginning in this section with the issue of coarse-graining.

Woodward maintains that the "variables of upper level causal theories are extremely coarse-grained from the point of view of fundamental physics"

(Woodward 2007, 80). Expressing a related idea, Hitchcock suggests that advanced theories of physics replace imprecise causal notions with precise mathematical relationships: "There are advanced stages in the study of certain phenomena when it becomes appropriate to eliminate causal talk in favor of mathematical relationships (or other more precise characterizations)" (Hitchcock 2007, 56). Claims such as these suggest the following instance of the anti-causal argument scheme:

2.1 It is an essential component of causal relations that they relate a small number of coarse-grained and imprecisely defined events to one another.

2.2 The dynamical models of our well-established theories relate to one another states that are precisely defined over entire time slices.

2.3 Causal notions do not play a legitimate role in any well-established theory of physics.

How good is this argument? Premises 2.1 and 2.2 state that there is a contrast between imprecisely defined and coarse-grained causal relata, on the one hand, and the input state of dynamical laws characterized in terms of the exact values of certain variables, on the other. Yet this contrast is in danger of confusing relations between the variables within a theory's models with the relation between a model and the physical systems it is meant to model. Both in the case of a theory's putatively causal claims and in the case of representing a phenomenon with the help of a physical theory's dynamical laws, we need to distinguish carefully the *models* constructed with the help of the theory from *the real-world systems* the models are intended to represent. *Within* a dynamical model, a theory's basic equations relate states precisely defined in terms of the values of certain state variables to one another. But within a causal model, the causal relations between variables can be similarly precisely defined, as formal approaches, such as the structural account of causation developed by Judea Pearl (Pearl 2009 [2000]) and the causal Bayes nets approach developed by researchers at Carnegie Mellon (Spirtes, Glymour, and Scheines 2000) show. Thus, both causal and purely dynamical models can contain formally precise relations between precisely defined states of the model.

By contrast, in the case of both causal models and purely dynamical state-space models *the relation between a model and the real-world system it is meant to represent* is to some degree vague, imprecise, and approximate. All scientific theories represent the phenomena in their domain only within certain limits of accuracy and with the help of abstractions and idealizations: neither causal nor purely dynamical modeling provide us with "perfect models" of the phenomena.

Third, we can ask about the relationship between models constructed with the help of different theories at different levels of grain. Woodward maintains, crediting Hitchcock for this point, that in the case of causal models the "coarse grained variables may fail to completely partition the full possibility space from the point of view of an underlying fine-grained micro theory" (Woodward 2007, 81). What Woodward has in mind here is the fact that from the perspective of an underlying micro theory, there may be states that are intermediate between two distinct macro states such that it may be unclear or vague which macro state these intermediate states realize or instantiate. Making a related claim, both Loewer and Field emphasize that quantities in the special sciences are multiply realized. Thus, Field takes it to be "fairly significant" that causal variables are inexact and defines that "a variable is inexact if the claim that it assumes a given value on an occasion can be realized in many different ways that on a deeper level of analysis are importantly different" (Field 2003, 445).

But once again this feature does not constitute a difference between causal models and the dynamical models of a physical theory, since the latter also fail to completely partition the possibility space of any underlying even more fine-grained theory, at least in the case of all but the most fundamental theory of physics – such as perhaps a final theory of quantum gravity.[1] According to one standard conception, physics provides us with a hierarchy of "effective theories" – that is, theories that adequately represent the phenomena at a certain length scale but break down at shorter lengths.[2] Each effective theory is coarse-grained with respect to theories of smaller length scales. Consider one of Woodward's own examples of a putatively fundamental theory – classical electromagnetism. Like models in the special sciences, this theory fails to partition the full possibility space from the point of view of an underlying quantum theory. Even though specifying the state of an electromagnetic system may require specifying the charge densities and field strengths at each spacetime point within some region, possible classical states do not completely partition the possibility space of an underlying quantum field theory, which, for example, permits of superposition states that have no classical analogue.

Thus, as far as the issue of *coarse-graining* is concerned, there is no difference in kind between causal models and purely dynamical physical

[1] It is perhaps not without irony that one of the main approaches to quantum gravity, the causal set approach, posits causal relations as the most fundamental relation, from which spatiotemporal relations are emergent (see, e.g., Rideout and Sorkin 1999).

[2] The term "effective theory" has been introduced in field theory, where an "effective field theory" is a field theory that, for a certain length or energy scale, ignores substructures or degrees of freedom that do not play an appreciable role at the length scale in question.

models – not for relations within a model, nor for model-world relations, nor for intertheoretic relations.

Moreover, it is unclear why we should endorse premise (2.1) and accept that causal representations have to be coarse-grained qua being causal. Both Field and Woodward sketch an argument for this claim, suggesting that certain asymmetries among correlations that exist among variables in causal structures crucially depend on the fact that these variables are coarse-grained from the perspective of a second, more fine-grained structure. The argument is this. A central assumption in drawing causal inferences from statistical data is the causal Markov condition (CM), which states that a variable A in a given causal structure is probabilistically independent of its non-effects B_j conditional on the set of its direct causes $\{C_i\}$. If we add the assumption that effects, at least in the kind of circumstances we are familiar with, do not precede their causes, we arrive at a temporal asymmetry that plays an important role in causal inferences: in trying to explain correlations between variables that are not related as cause and effect, we search for past common causes that screen off the correlation. By contrast, conditioning on the joint effect of two causes will generally render them dependent. Conditional on a common cause, two variables that are marginally dependent become independent, whereas conditional on a common effect, two marginally independent variables become dependent. This asymmetry between conditioning on common causes and on common effects arguably is a central component of the utility of causal notions, yet it seems to be in tension with the assumption of determinism.

Under determinism, if there is an event C in the past of two events A and B that screens A and B off from each other, then there will also be an event C^* that occurs after A and B and renders the two events conditionally independent. Here is an argument for this claim (see Arntzenius 1992). According to determinism, two worlds that are in the same state at one time are in the same state at all times. Now, for every event E at time t, there is a unique set of states $\{S_i\}$ at t, such that E occurs exactly if the world is one of these states S_i. Under determinism each of these states S_i will evolve into a unique state S_i' at some later time t'. For any set of states at one time, there is a unique set of states at any other time. If we also assume that there is an event that corresponds to each set of states – the event that occurs exactly if the world is in one of the states in the set – then it follows that for any two events for which there exists a screening-off event at one time, there will also exist a screening-off event at any other time. In particular, there will also be screening-off events at all times *after* the occurrence of the correlated events in question. Thus, given determinism, there is no screening-off asymmetry: for two events there exists an event in their past that renders them conditionally independent exactly if there exists an event in their future that renders them independent.

A common response to this argument is that the future screening-off event will be in general highly non-natural and non-localized: as Woodward puts it, "it will be very diffuse, spread out and gerrymandered"; or, as Field says, the variables representing the events in question have to be ones we find "salient." Thus, the screening-off asymmetry can be rescued, if we insist that the screening-off event be natural and localized. So far so good, but both Woodward and Field go on to suggest that the naturalness condition is at bottom a coarse-graining condition: Woodward says that "the asymmetry is in part a product of the particular coarse-graining of the macroscopic world that we adopt" (Woodward 2007, 90), and Field maintains that "with 'exact' variables . . . the asymmetry completely disappears in classical physics" (Field 2003, 446). Yet this final step in the argument is mistaken. The asymmetry does not depend on the assumption that the variables in question are coarse-grained or inexact in the sense of being multiply realizable by variables on a more fine-grained level; and the asymmetry does not necessarily disappear as we move to a more fine-grained representation. If anything, the opposite is true.

Field's and Woodward's discussions conflates two distinct issues. The first is whether the assumption of determinism is in tension with a time-asymmetric screening-off condition; the second issue is how the causal properties we attribute to a system might be affected by the choice of grain in the causal structure representing the system. *At a given level of grain* at which we can represent a range of phenomena with the help of deterministic laws of evolution, there may exist a set of salient, localized variables that is "causally privileged" over a set of non-salient, non-localized variables, even though the two sets of variables are interdefinable. But this has nothing to do with the question whether the variables at issue are exact in Field's or coarse-grained in Woodward's sense: a distinction between "natural" or "salient" variables and "unnatural" variables can be drawn at a given level of grain.

Now, according to David Lewis's sense of "natural" (which we briefly discussed in the previous chapter), the notion of naturalness allows for comparisons across different levels of grain. According to Lewis, multiply realized properties are less natural than their realizers. But this seems to suggest that the more closely we approach ever more fine-grained descriptions, the more "natural" the set of natural properties at that level of grain will be, and hence, the more robust the distinction between natural and unnatural properties will be at that level. Thus, moving to a more fine-grained level does not make the asymmetry disappear, as long as we make sure that the more fine-grained variables are also localized and "natural."

Moreover, the causal Markov condition and a common cause principle are provably true for deterministic systems with independently distributed exogenous variables (see Spirtes, Glymour, and Scheines 2000; Pearl 2009). A proof by Frank Arntzenius Arntzenius (1992) makes the application of common-cause reasoning to systems in physics particularly perspicuous and shows that a common-cause principle necessarily holds for deterministic physical systems whose initial states are independent of one another. If we assume a large number of pairs of systems $<A,B>$, such that the two members A and B of each pair have been isolated before a time t when they both interact with a third system C, then it follows from the assumption that the initial states of the two systems are statistically independent that any correlation between A and B has a common cause in C that screens off the correlation. Thus, the common-cause principle follows from the time-asymmetric assumption that the initial states of systems isolated from one another in the past are statistically independent of one another. As we have seen, it follows from the argument from determinism that there will also be a future screener off. But in general this screening-off condition will not be a property of the system C alone, but will be some non-natural joint "property" of all three systems.

In a bit more detail, the proof goes as follows. Let us posit that three systems A, B, C evolve deterministically in isolation from one another prior to time t. And let us assume that C interacts with both A and B at t such that the state of A after the interaction is determined by the states of A and C just prior to the interaction and that the state of B after the interaction at t is determined by the states of B and C just prior to the interaction. Further posit a probability distribution over the initial states of A, B, and C at some initial time t_i prior to the interaction such that the initial states are probabilistically independent. Then the initial probability distribution induces a probability distribution over the states of A, B, and C at all later times. It then follows that any probabilistic correlation between the states of A and B after their interaction with C is screened off by an earlier state of C.

The state of a system at a given time is given by the values that the variables representing the system take on at that time. The state can be represented as a point in the system's state space defined by the set of variables (such as particle positions) and their possible values. Let us assume that the variables in question are "exact" in Field's sense and that we take these variables to provide the most fine-grained representation available. Determinism implies that values of A's state variables after t are determined by the values of A and C at t_0. But let us further assume that the system is also deterministic at some higher level of grain and that there is some partition P of the state spaces of A and C such that for each pair of cells in that partition $<A_i; C_j'>$, the probability that A is in some final state a_f at a later time t_f is either 1 or 0. Similarly, let us assume that there is some partition P' of the state spaces of B and C such that for each pair of cells in that partition $<B_k; C_j''>$, the probability that B is in some final state b_f at a later time t_f is either 1 or 0. Then the system will be deterministic for all partitions more fine-grained than P and P' but not necessarily deterministic for partitions more coarse-grained than P or P'. In particular, the system will also be deterministic for the partition C_j that is the result of using C_j'' to further partition C_j'.

The proof then is as follows:

$$P(a_f \& b_f / C_i)$$
$$= \sum_{j,m} P(A_j \& B_m) P(a_f \& b_f / A_j \& B_m \& C_i) \quad \text{from the probability}$$
$$\text{calculus.}$$

$$P(A_j \& B_m) = P(A_j) P(B_m) \quad \text{initial independence.}$$

$$P(a_f \& b_f / A_j \& B_m \& C_i) = P(a_f / A_j \& B_m \& C_i) P(b_f / A_j \& B_m \& C_i)$$
$$= P(a_f / A_j \& C_i) P(b_f / B_m \& C_i) \quad \text{from the determinism assumption.}$$

Thus:

$$P(a_f \& b_f / C_i) = \sum_{j,m} P(A_j) P(B_m) P(a_f / A_j \& C_i) P(b_f / B_m \& C_i).$$

And therefore:

$$P(a_f \& b_f / C_i) = P(a_f / C_i) P(b_f / C_i) \quad \text{from the probability calculus.}$$

If we assume that the interactions are governed by a relativistic theory and that A and B are spacelike separated for the entire period from t_i to t_f, then a_f is determined by the cross section of its backward lightcone at t_i, which contains only a_i and c_i, but not b_i, whereas c_f is determined only by the full state of a_i, b_i, and c_i. If the system is governed by both forward and backward deterministic laws, then similarly a_i is determined by the cross section of its forward lightcone at t_f, which contains a_f and c_f but not b_f, and c_i is determined by the joint of a_f, b_f, and c_f. Thus, purely dynamically there is no asymmetry. The asymmetry is introduced by the assumption of initial independence, which makes common-cause reasoning possible, adding a significant inferential resource to purely dynamical inferences based on the dynamical laws and initial or final states. Although dynamically c_i is determined only by a full specification of the joint state a_f, b_f, and c_f, the principle of the common cause allows us to infer the presence of a common cause c_i purely from the knowledge of correlations between a_f and b_f. Thus, causal representations allow us to make inferences that are not possible within a purely dynamical non-causal model. I will discuss this point in much more detail in Chapter 5.

Field's characterization of the natural set of properties as "salient" may suggest that the common cause asymmetry is in some sense subjective and dependent on what kind of properties we as human observers may find particularly interesting. But, at least in the present context, the assumption of probabilistic independence has a non-subjective core. If we consider systems that are isolated from on another both prior to and subsequent to an interaction at time t, then the independence assumption states that the localized states of the individual systems are probabilistically independent *before* to the interaction but in general *not after* the interaction.

Let us assume that the premises of Arntzenius's proof are satisfied for a given partition at a given level of grain. What happens as we move to different levels of grain and introduce either more coarse-grained or more fine-grained partitions of the systems' state space? The screening-off asymmetry might disappear if we move to a more *coarse-grained* level, as Arntzenius (1992) shows. If we allow that macroscopic events can have microscopic causes, then a screening-off condition that holds at a certain level of grain may fail to hold for more coarse-grained models. The reason for this is that the common-cause principle does not hold for a class of events that have causes outside that class. As we move to a more coarse-grained partition, the assumption of determinism might no longer be satisfied. Indeed, if one assumes that microscopic events can have macroscopic effects, then one might think that – turning Woodward and Field's conclusion on its head – the common-cause principle could not hold for macroscopic events.[3] Thus, Arntzenius concludes that "the only class of events for which the common

---

[3] Thus, it is somewhat surprising that Field refers to this very argument by Arntzenius as providing support for his conclusion that the asymmetry disappears at the level of the exact variables in physics. That there might be no screening-off cause among a set of coarse-grained variables, even though one exists for a representation of a given phenomenon in terms of more fine-grained variables, is also emphasized by Woodward and Hausman in their discussion of Salmon's example of the correlation between the motion of two billiard balls after they were struck by the cue ball: "there will be no screening-off common cause if one is confined to coarse-grained variables such as 'collision' or 'no collision.' It is only at a more refined level of description that one will be able find a screener off" (Hausman and Woodward 1999, 529).

cause principle could possibly hold is the entire class of microscopic events"
(see Arntzenius 1992, 230). Yet Arntzenius's conclusion is too strong: even if
macroscopic events can and do have microscopic causes, there may be many
circumstances in which it is appropriate to ignore these causes and represent a phenomenon in terms of a complete and Markovian macroscopic
causal model.

Field, as we have seen, claims that "with 'exact' variables . . . the asymmetry completely disappears in classical physics." Since he assumes determinism at that level, this means that he is committed to the idea that
the assumption of initial probabilistic independence fails as we move to
a more fine-grained partition. But this commitment is implausible, and
Field offers no argument for why we should accept it. For probabilistic
dependencies that are absent for a more coarse-grained partition of the
systems' state space to be introduced at the level of a finer-grained partition
requires that the choice of fine-grained partition delicately depend on the
probabilities in ways that one would not expect to be true in general. Let
us assume that the independence assumption holds for the cells A_j and B_m
of a certain coarse-grained partition:

$$P(A_j \,\&\, B_m) = P(A_j)P(B_m).$$

Then, let us assume that (to keep things simple) we divide each cell A_j
further into two cells A_x, A_y and do not further partition the cells B_m:

$$P\big((A_x \vee A_y) \,\&\, B_m\big) = P(A_x \vee A_y)P(B_m),$$

which, with the help of Bayes's theorem and the principle of total probability, can be rewritten as:

$$\big((P(B_m|A_x) - P(B_m)))P(A_x) = \big((P(B_m - P(B_m|A_y))))P(A_y).$$

This equality holds in general, if $P(B_m|A_x) = P(B_m)$, and similarly
for A_y – that is, if the independence assumption continues to hold for
the more fine-grained partition $A_{x,y}$. For it to fail, the choice of partition would have to be delicately matched to the conditional probabilities
$P(B_m|A_x)$ – something we would not expect to be true in general.

Consequently, contrary to Field's claim, an initial independence assumption is frequently made in physics even for micro states. Thus, Oliver Penrose and Ian Percival use such an assumption on the micro-physical level to
derive a screening-off asymmetry similar to the one derived by Arntzenius
(Penrose and Percival 1962). A similar assumption also plays an important
role in the foundations of thermodynamics, as we will see in Chapter 8.

In this section I have argued, first, that causal models are coarse-grained in the very same ways in which any scientific representation, including representations in physics, are coarse-grained; and, second, that being coarse-grained in the sense of being multiply realized is not a necessary condition for a causal representation. That a deterministic causal model satisfies the causal Markov condition follows from the fact that the exogenous variables are statistically independent.

3. Complete models and background conditions

A second putative contrast between causal models and models in fundamental physics concerns the role of *background conditions*. The models of fundamental physics, according to Field and Woodward, provide us with complete representations of a phenomenon that include all possible influences on the phenomenon in question. Consider once more the example of the lighting of a match. Within the context of a relativistic physical theory, it would seem, the entire state of a region of the world prior to the lighting would have to be specified in order to be able to derive that the match will light, without allowing us to draw a distinction between salient causes and background conditions. In fact, if we assume a relativistic theory, the entire backward lightcone of an event E threatens to come out as a cause of E. But this, it may seem, is an absurd conclusion – clearly (one might say) the striking of the match is a cause of the lighting in a way in which the state of the room next door is not – and in order to resist it we may want to conclude that the concept of cause is essentially a concept meant to express the kind of dependencies between small numbers of variables relative to background conditions, with which the special sciences or common sense are concerned.

This suggests the following instance of the general argument schema:

3.1 If a model does not permit a distinction between causes and background conditions, then the model cannot be interpreted causally.

3.2 Physical theories provide us with complete models of the phenomena in their domain, constructed from the data on complete lightcone cross sections as input, which do not permit a distinction between causes and background conditions.

3.3 Therefore, there is no place for causal notions in physical theories.

One response to this argument is to challenge 3.1 and to argue that what we learn from models of relativistic theories which take into account entire cross sections of backward lightcones is that an event has many more causes than we might have naïvely assumed. "What's the big deal in that?" Field himself asks rhetorically, after presenting a version of this argument. Yet he immediately replies that this response is unsatisfactory, since (in terms of his own example involving a woman aiming a hose at a fire and a man praying that the fire would go out)

> there would be a big deal if we had to conclude that if c_1 and c_2 are both in the past light cone of e then there is no way of regarding one of them as any more a cause of e than the other: then Sam's praying that the fire would go out would be no less a cause than Sara's aiming the water-hose at it, and the notion of causation would lose its whole point. (Field 2003, 439)

Accepting all events in the backward lightcone of an event as cause, Field maintains, would eviscerate our notion of cause.

I will examine Field's claim later, but first I want to focus on premise 3.2 in the argument. As I stated it, the argument is not valid, at least on one reading of 3.2. If not all models in physics were complete lightcone models, then we could at most establish that these models could not be interpreted causally, but not that there is no room for causal notions in physics. In order to render the argument valid, 3.2 has to be disambiguated as follows:

> 3.2* *All* representations with which a physical theory provides us are complete models of the phenomena in their domain in that they involve the complete specification of initial states.

Yet (3.2*) is wrong for at least the following two reasons.

First, characterized at the most general level, solving the equations for a system requires solving a boundary value problem. The boundaries in question can be *both* temporal *and* spatial boundaries. Initial (or final) conditions specify the state of the system at one time – that is, on one time slice or on a spacelike hypersurface. Initial conditions are *temporal* "boundary conditions." Sometimes initial conditions are distinguished from "proper" boundary conditions that specify the *spatial* boundaries of a system across time. Boundary conditions might specify the state of the walls of a container, the state of the endpoints of a vibrating string, or, in the case of a mechanical system, the satisfaction of certain constraints. A solution to the dynamical equations represents how a system with specific initial and boundary conditions evolves in time. A pure initial-value problem of the kind Field imagines is merely a special type of boundary problem and whether such boundary conditions are appropriate depends on the type of equation at issue (see, e.g., Snider 1999, 262–7).

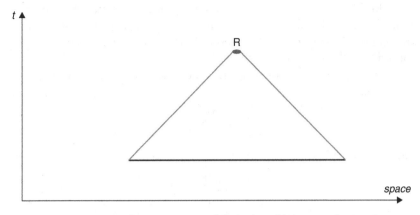

Figure 3.1 A complete cross section of the backward lightcone of region *R*

So-called *hyperbolic equations*, like the wave equation, which in two dimensions is

$$\frac{\partial^2 \psi}{\partial x^2} - \frac{\partial^2 \psi}{\partial y^2} = 0,$$

can be solved as a pure initial- (or final-) value problem. I will discuss this equation and possible solutions to it in detail in Chapter 7. Yet in many physically realistic circumstances, even the wave equation has to be solved with the help of a mixed initial- and boundary-value problem. Hyperbolic equations imply a finite propagation velocity. Thus, in the case imagined by Field, that of a relativistic system, initial conditions have to be specified on an entire cross section of the backward lightcone of the spatial region occupied by the system at the time of interest – that is, that region from which influences could reach the point of interest by traveling at most at the speed of light (see Figure 3.1). In this case, no spatial boundary conditions are needed. If no influence can travel faster than the speed of light, then nothing occurring outside of the past lightcone can affect the state of the system at *t*. But the vast majority of models of actual physical systems are not constructed with the help of a pure relativistic initial-value problem and include a specification of boundary conditions as well. After all, if we wanted to model the evolution of a system during a time period even as short as 1 second, the cross section of the backward lightcone on which we would have to specify the relevant initial data would have a diameter of 300,000 km. Thus, the range applicability of pure initial value problems is rather limited. As Arthur Snider says, in the context of discussing the one-dimensional wave equation for a string, "unfortunately the initial-value problem for the string seldom corresponds to physical reality. The string is not infinitely long" (Snider 1999, 234).

So-called parabolic equations, such as the heat equation, which in two dimensions is

$$\frac{\partial^2 \psi}{\partial x^2} - \frac{\partial \psi}{\partial y} = 0,$$

and the *elliptical* Laplace equation, which in two dimensions is

$$\frac{\partial^2 \psi}{\partial x^2} + \frac{\partial^2 \psi}{\partial y^2} = 0,$$

cannot in general be solved through a pure initial-value problem and typically require spatial boundary conditions as well. Elliptical equations require boundary conditions where either ψ or $\partial \psi / \partial n$ are specified on the entire boundary surrounding the solution region; parabolic equations require initial plus boundary conditions (see, e.g., Snider 1999, 265). Thus, Field's characterization of how physical theories model the world fits at most one specific type of equation that we commonly find in physics. The toolbox of physics contains a far richer set of tools than just the hyperbolic equations to which Field's discussion applies (and even there it applies only with important qualifications).

How does the fact that models may include spatial boundary conditions affect the argument just presented? Arguably, boundary conditions can play a role analogous to that of background conditions in causal models in the higher sciences: an arrangement of gears in a mechanical model provides the background conditions under which certain dynamical relations hold, just as the presence of oxygen is a background condition for the lighting of the flame. And that certain quantities are zero on a spatial boundary may be taken to be a shielding condition similar to one that allows us to exclude the state of the room next door in evaluating the causal claim that striking the match causes a flame.

3.2* is false for a second reason. Very often, even within the context of established or putatively "fundamental" theories, we represent phenomena not in terms of a complete initial- and boundary-value problem, which would provide us with information about the precise physical state of each spacetime point in the spacetime region of interest, but by specifying only a finite number of relevant components of a system at different levels of grain and then showing how these components interact. Although the use of differential equations may constitute progress (as Field and Hitch-cock suggest), this progress has not resulted in the general replacement of incomplete and partial models with complete lightcone models. I argued in Chapter 2 that *all* scientific representations are partial and distorted, but for present purposes it is enough to show that there are *some* models constructed within the context of mature or fundamental physical theories that are incomplete and partial. After all, the overall thesis of this book is not that every model in physics is causal but that causal representations play an important role in physics.

As a case study, to which I will return repeatedly in this chapter and the next, consider how the proton beam in the Large Hadron Collider (LHC) at CERN is modeled (see, e.g., Steinhagen 2007). I choose this particular example because it is as good an example as any of modeling in physics involving fundamental theories, in this case classical electrodynamics, quantum mechanics, and, of course, the standard model of particle physics, which unifies the electromagnetic, weak, and strong interactions. The LHC is a circular accelerator in which two proton beams are accelerated and then brought to collide with each other. Some components essential to its operation are bending magnets that are responsible for keeping the beams on their circular orbit around the LHC ring, focusing magnets that prevent the beams from spreading and focus the particle orbits at the center of the LHC vacuum chamber, and accelerating systems that accelerate the protons. Measurements of the collision products are intended to provide useful tests of the Standard Model of particle physics.

What is important for our purposes here is that the proton beams in the accelerator and the decay products in the particle detectors are not represented in terms of a complete quantum field theoretic model constructed with the help of the Standard Model – that is, in terms of a model, including the quantum field theoretic micro state of the world in an entire cross section of the backward lightcone of the protons' orbits. Rather, the model of the proton beam is partial, ignores many influences on the accelerator and detector that should in principle be present, and treats different components of the experiment in terms of different theories at different levels of grain. Interactions between the proton beam and the accelerator are treated largely using the resources of classical electrodynamics – an established theory of physics, but one that is considerably more coarse-grained than the Standard Model. The proton beam is taken to interact directly only with the highly localized electromagnetic fields produced in the various components of the accelerator, such as bending and focusing magnets. The different pieces of hardware included in the LHC are modeled macroscopically – the precise states of all the subatomic particles that make up, for example, the bending magnets is not part of the model, and there is no full quantum-field theoretic model of the detection chambers. External environmental influences on the accelerator are also modeled, both qualitatively and quantitatively, but physicists are selective in deciding which external perturbations need to be included in a model of the proton beam, and all such perturbations are modeled at an extremely coarse-grained level from the perspective of the more fundamental theories at issue. For example, motions of the ground surrounding the accelerator are included in the model by representing the accelerator as being embedded in a homogeneous viscous medium that exhibits both random and coherent motion, which is modeled as plane waves propagating in the medium. These models are compared with seismic measurements. The effects of tidal forces on the accelerator geometry are modeled as well, yet the micro state of a pot of cheese fondue in a nearby mountain restaurant is not included in the model.

Not only what happens in the accelerator but the detection events, too, are not modeled in terms of a complete quantum field theoretical model, which would be absurdly and impossibly complex. In fact, the only events that are modeled in terms of the Standard Model itself are the various production and decay channels of Higgs bosons and other elementary particles as a result of the proton-proton collisions. For example, the proton collisions can result in the creation of vector boson pairs, either W or Z bosons, which can fuse into a Higgs boson: WW → H or ZZ → H. The Higgs boson can decay into two photons: H → $\gamma\gamma$. It is *only* these interactions at the core of the CERN experiment that are modeled with the help of the Standard Model. The particle detectors and possible detection events are, like the particle accelerator, modeled at a much more coarse-grained level. The Higgs boson is detected indirectly through detections of its decay products. Particle detectors function by measuring energies and momenta of the various decay products. One example of such a detector found in the ATLAS experiment at CERN is a semiconductor tracker consisting of silicon microstrips, which measure the trajectory of a particle through the locations of electrons that are knocked from the atoms in the material. This results in the production of "electron-hole pairs," which migrate through the silicon strip and are what is ultimately detected. The physics used to model the detector processes is solid-state and semiconductor physics, a theory that from the perspective of the Standard Model is extremely coarse-grained.

The lesson I want to draw from this discussion is that even the models physicists construct to represent experiments aimed at testing our most fundamental physical theories are partial and involve different levels of coarse-graining and a distinction between relevant factors and background conditions. Of course, there are scientifically legitimate reasons for this treatment: the proton beam is sufficiently well shielded from may external influences that these can be ignored in the model; the precise state of the atoms in the bending magnets or of the surrounding mountains does not (within limits) appreciably affect the total magnetic field and hence can be treated in a coarse-grained manner; and the energies of the particles composing the detector are to low to fall within the domain of quantum field theory. There are, then, good reasons for ignoring many of the perhaps in principle existing influences on the protons' trajectories and on the collision products; there are also good scientific reasons for treating some of the influences in what from the perspective of the Standard Model is an extremely coarse-grained way. But the very same kinds of reason allow us to use partial and coarse-grained representations in the construction of paradigmatically causal models in the special sciences or in commonsense reasoning. The state of the neighboring room can be ignored in our causal model of the lighting of the match, since the wall between the rooms sufficiently shields the match from any potential influences on the lighting resulting from changes to the state of the neighboring room. And the precise microscopic state of the match (within limits) will not affect whether or not it lights.

Thus, again, we have not been able to identify a genuine difference between modeling in physics and causal modeling elsewhere. Modeling in fundamental physics, too, proceeds by representing explicitly only a limited range of factors in a model. Some of the included factors are treated at different levels of grain, while other factors that possibly might have an influence but are in the circumstances at issue well shielded are relegated to the status of only vaguely specified background conditions. In both dynamical and causal models, background factors do not show up within the model but are appealed to as part of the justification why a given model is appropriate in the circumstances at issue. If anything, these features characterize modeling in fundamental physics even more fully than modeling in other sciences, and one might be tempted to conclude that the more fundamental a theory is, the more restricted is its domain of genuine modeling applications. Once we have reached the level of quantum field theory, the *only* events that are modeled fully with the theory are the productions and decay events of elementary particles at the experiment's core, which take place in particle detectors that themselves are modeled at a much more coarse-grained level.

As somewhat of an aside as far as the present argument is concerned, it is worth pointing out that the ways in which the functioning of the LHC is described even in the scientific literature support Cartwright's thesis that even at the level of fundamental research in physics, our conception of the world is ineliminably causal. For example, in a report from the LHC study group of CERN on the LHC design (Pettersson and Lefèvre 1995) we learn that there are various places in the machine where beams can be "injected," that other components allow "suppression" of dispersion, and that others allow for the "cleanup" of the beam. Finally, there is the "beam dump" where the beam can be deposited with the help of "kickers." (Thus, the report informs us that "the principle of the beam dumping system is to horizontally kick the circulating beam into an iron septum magnet.") In the detector, when a photon passes through matter, it "knocks out" electrons from the atoms "disturbing the structure of the material" and "creating" loose electrons. Although the word "cause" is not used in these descriptions, the terms I quoted all describe what Cartwright would characterize as "concretely fitted out" instances of "causings," and, hence, one gets the impression that the operation of the LHC – from the injection of proton beams, to their collision in the detectors or their being dumped in the beam dump, to the detection of the decay products – is a causal process through and through.

Field maintains that the problem with putatively causal models of a physical phenomenon would be that "facts about each part of the past light cone of an event are among the causes of the event." But the models physicists in fact *use* to represent actual phenomena very rarely if ever are the complete structures that philosophers postulate, because these structures are in general much too complex for us ever actually to possess. Yet the discussion so far generates a puzzle: How can it be that physicists manage to construct such partial and idealized models given the feature of differential equations to which Field draws our attention? How can it be that models in physics are partial, if the relevant dynamical equations require a specification of the precise data for each spacetime point on complete initial and boundary surfaces as input? To solve an initial-value problem, we need to be able to specify the complete state on an appropriate initial-value surface. Yet this state would be much too complex, and if we wanted to solve a pure initial-value problem in the case of a hyperbolic equation, the entire cross section of a backward lightcone would (in most cases) be much too large.

A discussion of the various strategies by which actual models avoid the unwieldy complexities of complete initial-value problems over cross sections of entire lightcones would take us too far afield, but among the strategies are the following. First, as we have seen, models may include boundary conditions that represent facts about shielding conditions and which allow us to severely restrict the spatiotemporal region that must be represented in a model. Without such restrictions, representing actual phenomena and confirming the adequacy of our representations would be impossible.

Second, models may set the values of variables identically equal to zero in large subregions of the initial-value surface or may not even contain variables representing possible influences arising from large regions of the initial-value surface, if whatever is physically going in these regions can be taken to be irrelevant to the phenomenon at issue. For example, the interaction of two billiard balls on a billiard table can be treated as a constrained two-body problem, without containing any additional variables that represent additional influences on the motion of the two balls due to other bodies or fields. That is, we represent the interaction between the two billiard balls as if it were taking place in a simple universe that contained nothing but the two billiard balls and the table and a gravitational field constraining the balls to the surface of the table.

Third, it is important to recall the distinction between models and the real-world systems they are intended to represent: In some sense the viscous fluid model of the ground surrounding the LHC at CERN is extremely coarse-grained and partial, since it leaves out many of the details characterizing the rock formations in which the accelerator was built. But nevertheless the model contains a specification of the values of certain quantities at each spacetime point. That is, it is important to distinguish two senses in which a model may be thought to be complete: in order to bring the apparatus on initial or boundary values to bear, a model has to be *mathematically* complete in the sense that it has no "holes" or "gaps" on the initial or boundary surfaces for which the value of the model's variables remain unspecified; yet this does not meant that a model needs to be *physically* complete and represent completely and precisely the physical system modeled.

Fourth, models need not represent a phenomenon in terms of a full-fledged initial-value problem. For example, scattering phenomena are often modeled as interactions between a scattered particle and a scatterer (and sometimes as an interaction between a microscopic particle and a macroscopic medium) without specifying a complete initial-value problem, which would require a specification of the initial fields and of the microscopic states of any other charged particles in the vicinity. That we can model a phenomenon without solving a full-fledged initial-value problem will be important in later chapters. In particular, I will argue that an important formal tool for representing causal relations is provided by the so-called Green's function associated with an equation, which specifies how a localized pointlike disturbance propagates through a system.

To conclude the discussion in this subsection so far, I have argued that the practice of modeling in physics shares crucial similarities with causal modeling elsewhere, and thus we should reject premise 3.2* in the argument given earlier.

I now want to turn to premise 3.1 (which states that if a model does not permit a distinction between causes and background conditions, then the model cannot be interpreted causally) and argue that this premise should be rejected as well. The reason is that we can grant Field and Woodward's contention that the distinction between salient causal factors and background conditions plays an important role both in ordinary discourse and in the special sciences and yet resist the claim that it is *essential* to the notion of cause as difference-maker.

The putatively complete models to which Field draws our attention do not allow us to draw a principled distinction between cause and background condition. Yet that in itself does not make causal notions inapplicable to such models, if we insist that this distinction is an important *pragmatic* component of our concept of cause, which reflects the close connection between the concepts of cause and of explanation. Not everything that in principle can make a difference to the occurrence of a particular event is salient in every context: in some context the striking of the match is the factor on which we wish to focus, but we can imagine other contexts in which the presence of oxygen becomes the focus of our attention. Yet in addition to its pragmatic dimensions, we can also identify a less pragmatic minimal core of an asymmetric concept of cause, which includes any and all factors that could in principle make a difference to the occurrence of an effect. Moreover, that the distinction between causes and background conditions is heavily pragmatic does not mean that it is merely subjective or arbitrary. Rather, there may be very good "objective" reasons for excluding a certain factor from the set of possible causes of an event in a given modeling context.

As we have seen, Field worries that once the entire backward lightcone of an event e is included in the set of its causes, then we have to treat all macro events in the past of e as causally on a par: Sam's praying comes out as a cause of the fire's going out just like Sara's aiming the water hose at the fire. Woodward discusses another example in this context, which I want to use here to illustrate and examine Field's claim. Suppose that the event e is the event of my headache disappearing (represented by the variable E); that 30 minutes prior to that event (at t) I took aspirin (event a represented by variable A); that around the same time t my neighbor sneezed (event s represented by S); and that I wished my headache would disappear (event w represented by W). Whereas taking aspirin may be an effective strategy for making a headache go away, bringing about events w or s is not. Yet, Woodward points out, if we consider very fine-grained microscopic specifications (E^*, A^*, W^*, S^*) of the states of the fundamental particles realizing events e, a, w, and s, respectively, then it will turn out that E^* does not only depend on an exact specification of A^*, but also on a specification of W^* and S^*, and in fact of the entire cross section of the backward lightcone at t. According to Woodward, this threatens to undermine the whole point of the notion of causation. "If we are forced," he says, "to the conclusion that [sneezing and wishing] are causes of e as well, the motivation (according to manipulationist accounts) for introducing the notion of causation in the first place – the intuitive connection between causation and manipulation and the contrast between causal and merely correlational relationships – appears lost" (Woodward 2007, 84–5). But is the antecedent true – are we indeed forced to the conclusion that sneezing and wishing are causes of the headache's disappearance, once we posit not only a macroscopic model but also a putatively complete micro model of the situation? The answer, as I want to argue now, is "no." For reasons that Woodward himself discusses in his paper (see Woodward 2007, 85–7), we can preserve the intuition that my neighbor's sneezing is not a cause of my headache's disappearing, even though the precise micro state of my neighbor during his sneezing might be a cause of what the precise micro state of my brain is as my headache vanishes.

There seem to be two related worries concerning a complete putatively causal micro model. First, one might worry that it simply follows from the fact that the *fine-grained* variables are causally related – that is, that S^* is a cause of E^* – that the *coarse-grained* variables are causally related as well. Second, the argument might be that once we grant that E^* depends on S^*, we have to worry about extreme "fillings" for the values of the variables S^* that would affect the occurrence of E. In both cases we are led to conclude that S is a cause of E, even though we do not think that we can manipulate E by means of manipulating S. Thus, if we want to retain the connection between causation and manipulation, we have to deny that there is fine-grained causal dependence: on the micro level there is nomic dependence but no causal dependence.

In reply to the first worry, we can concede that the *precise* disappearing of the headache E^* depends on the occurrence of W^* and S^*, without accepting that neighbors' sneezes cause headaches to disappear. Arguably a fine-grained causal model will imply that interventions into the micro variables involved in the characterization of S^* will result in changes to the micro variables involved in the characterization of E^*, yet the problematic causal claim is not this but the claim that the value of a *coarse-grained* variable representing sneezing is causally related to the value of a *coarse-grained* variable representing the disappearance of headaches. Presumably, however, changes in the values of the micro variables characterizing S^* that are associated with a change in the macro "sneezing variable" S from 1 to 0 will *not* result in changes in the values of the micro variables characterizing E^* that are associated with a change in the macro variable E. That is, while the *microscopically precise* way in which the neighbor sneezes will make a difference to the *microscopically precise* way in which my headache disappears, whether my neighbor sneezes or simply sits there quietly will not make a difference to whether my headache disappears. Thus, both on the micro level and on the coarse-grained macro level, the connection between causation and effective strategies for manipulation is preserved: the fine-grained causal model predicts that the precise way in which the sneezing occurs affects the precise way in which my headache disappears, and intervening into the precise way in which the sneeze occurs is an effective strategy for manipulating the precise manner of the disappearance of my headache. Yet since changes to how and even whether the sneeze occurs only affect the fine-grained details as to *how* my headache disappears, but not *that* it disappears, there is no causal relationship between the coarse-grained two-valued variables representing the occurrence of a sneeze and the disappearance of a headache. By the same token, intervening into whether a neighbor's sneeze is not an effective strategy for making headaches go away.

What about the second argument? The worry is that our discussion so far has ignored the possibility of there being extreme changes to the variables S^* that will result in my headache not disappearing. For example, we might assume that if my neighbor had emitted a sufficiently loud and shrieking noise instead of having sneezed, then my headache would not have disappeared. Nevertheless, we do not think that manipulating whether the neighbor sneezes can be part of an effective strategy for combating headaches, or so one might say.

Yet if we were to consider my neighbor's emitting a shrieking noise as a genuine possibility, then it seems that we ought to include in our coarse-grained causal model a variable that takes on different values whether the neighbor shrieks or emits less bothersome sounds, such as those associated with sneezing. Whether my neighbor sneezes rather than shrieks will make a difference to whether my headache disappears. Thus, the neighbor's sneeze would indeed play a causal role in my recovery from the headache and, once again, the relation between causal claims and claims about interventions is preserved. If we are considering that changes to the neighbor's sneezing can affect whether or not my headache disappears, then intervening into the neighbor's sneeze in the right way will be a way of influencing the disappearance of my headache.

In many contexts, however, we can have very good reasons for excluding "extreme fillings" – since we might know that my neighbor is not inclined to emit loud shrieks, or that the walls of his house are thick enough to block any but the very loudest shrieks – and in such contexts it is reasonable to adopt a causal model that allows only for a fine-grained dependence of E^* on S^* but not for a coarse-grained dependence of E on S. To some extent, of course, it is a pragmatic question about what possible values for the fine-grained or coarse-grained variables we are willing to consider in a given causal model. Yet this pragmatic dimension is part of the construction of causal models more generally: which particular causal model we choose in order to represent a certain situation is to some extent due to pragmatic and context-dependent factors. Moreover, as we have seen, similar context-dependent constraints also guide our construction of dynamical models of a given situation. The model of the rock surrounding the LHC as homogeneous viscous medium might not be able to correctly predict the behavior of the LHC if there was a very strong earthquake in Geneva, but there may be good reasons for ignoring this possibility in the model.

Thus, Field's argument that putative complete microphysical models of a phenomenon could not be interpreted causally, along the lines of a notion of causation as difference maker, fails. To sum up my discussion of the argument, let me return to the example mentioned by Field himself: Even though the *precise micro state* of Sam's body comes out as a cause of the *precise micro history* of the fire – as well it should – his praying is not a cause of the fire going out, as long as the relevant range of values of the variable representing his actions in our causal model includes only values representing his simply sitting there, performing a dance, or praying, and the like. Yet once we include "extreme" values for the variable representing Sam's actions, such as a value representing his dousing the flames with gasoline, then his praying does indeed come out as causally relevant to the fire going out and plausibly so.

4. Default behavior

We can distinguish two different concerns in the background of Field's or Woodward's worry that putatively causal models of established theories of physics would have to be fine-grained models that include the entire backward lightcone of an event among the event's causes. The first concern is that including too many of an event's causes in a causal model renders the notion of cause useless. If the entire backward lightcone is among an event's causes, then the notion of causation would lose its point, or so it may seem. The second concern is that, if the set of an event's causes becomes too large, there will be many causal claims inferable from a given model that conflict with central intuitions concerning the assertability of causal claims. Hence, Field's worry that a physical causal model of the history of Sara's putting out the fire would imply that Sam's praying is a cause of the fire as well.

I have argued that if we focus on how modeling in physics works – modeling even in the context of our more fundamental theories – these worries can be met. In constructing a model of a phenomenon or system, physicists make choices concerning appropriate levels of grain and boundary conditions and selectively choose which quantities to represent in what manner. In their paper "Actual Causation and the Art of Causal Modeling," Joseph Halpern and Chris Hitchcock discuss various decisions a modeler faces in constructing a causal model. As they argue, a modeler has considerable leeway in choosing the number of variables and their ranges, since "nature does not provide a uniquely correct set of variables" (Halpern and Hitchcock 2014), nevertheless there are context-dependent considerations that affect the range of choices that will be appropriate. As we have seen, the same is true of modeling in physics. Nature does not dictate which physical quantities we should model at which level of grain, and she determines neither which variables to include in the model nor what the correct ranges for the variables are. Even in the context of a full initial- or final-value problem for a relativistic theory, physicists can decide which physical quantities to represent explicitly on the backward lightcone of the event of interest. As I have argued, that a model constructed with the help of an initial-value problem needs to be complete mathematically does not entail that it also needs to be complete physically. Nevertheless, this does not render model construction completely arbitrary. Even if nature does not dictate a uniquely correct model, there can be perfectly good physical reasons for choosing certain variables and ranges in a given context rather than another.

One might worry, however, that these considerations do not go far enough in aligning the causal claims underwritten by putatively causal models with our core causal intuitions. Consider the following example to illustrate this worry. Imagine a radio antenna broadcasting a signal, which is received by a radio. We can model this situation with the help of an initial-value problem that represents the received signal as depending on the signal emitted by the antenna as its source. In keeping with what I argued earlier, we can decide which quantities to represent explicitly in our model in setting up an initial-value problem. Thus, it may be legitimate to treat the transmission of the signal as occurring in a "world" that contains only two objects: the antenna and the radio. If, however, we model the transmission of the signal with the help of a field theory, we have to specify the values of the electromagnetic field on the entire initial-value surface even if we take the field to be identically equal to zero on that surface. The overall strength and shape of the signal received by the radio depends, according to the model, both on the signal emitted by the antenna and on the strength and shape of the incoming field. If we were to treat this dependence causally, it seems that we would be forced to conclude that the incoming field's being equal to zero at some earlier time is just as much a cause of the playing of Beethoven's "Tempest Sonata" through the radio's speakers as is the emission of the broadcast signal by the antenna. And this conclusion may strike many as intuitively wrong.

A simple causal model of the situation would consist of three variables representing the received signal R, the emitted signal E, and the free incoming field F, respectively. R depends on both E and F symmetrically and in exactly the same way – the value of the received signal R is simply the superposition of F and E – and, hence, if one of the two is a cause of R, the other should be as well. Indeed, on any account of causation that evaluates the truth of causal claims in terms of some kind of counterfactual dependency (as a structural equation approach does), F is a cause of R. The problem is that in constructing a dynamical model through an initial-value problem, we have to include variables representing the initial field in the model and cannot relegate the initial field to the status of background conditions external to the model. But once a variable for the initial field is included in the model, we can ask how the signal received by the radio would change if the initial field took on non-actual values, making the conclusion that both initial fields and the antenna play equal causal rules difficult to avoid.

Consider the notion of "actual cause" introduced by Halpern and Pearl (2005). According to their account, a variable C is an actual cause of E exactly if E counterfactually depends on C under certain contingencies. Applied to our example, F is an actual cause of R exactly if it is the case that if we keep the value of E fixed to one of its values, there is some non-actual value for F such that R takes on a non-actual value. And indeed there are such values. For example, the incoming field might have the same frequency and strength as the broadcast signal but be perfectly out of phase with that signal so as to perfectly destructively interfere with the latter signal and exactly cancel it. Or the incoming signal might be such as to exactly destructively interfere with the signal broadcasting the "Tempest Sonata" and in addition carry a signal of a live performance of "Terrapin Station" by the Grateful Dead. Thus, that the incoming field is zero appears to come out as a cause not just of the radio receiving any broadcast signal but even of the radio playing a particular piece of music!

The present problem, however, is not restricted to modeling in physics and arises for any broadly counterfactual approach to causation, such as accounts analyzing causal notions in terms of structural equations, which quantify the causal dependencies among a set of variables. Halpern and Hitchcock (2014) call it the problem of isomorphism. The simple causal structure in our example is isomorphic to other structures in which it seems perfectly appropriate to treat both variables on which the effect depends[*] equally as causes. The lesson Halpern and Hitchcock (and others) draw from this problem is that there must be more to our judgments of causal relations than what can be captured in the causal structures defined by structural equations. This additional content can be captured in terms of a notion of 'normal' or 'default' behavior of a system. Halpern and Hitchcock propose that in assessing causal claims we consider only counterfactual worlds that are at least as normal as the actual world. Applied to our example, a delicately set up incoming field that is perfectly correlated with the radio signal and destructively interferes with the signal is less typical or normal than a zero incoming field. Since there are no counterfactual changes to the zero incoming field that take us to a world that is at least as normal as the actual world, the incoming field does not come out as an actual cause of the signal received by the radio. By contrast, the default state of the antenna arguably is a "standby" state in which it does not transmit any signal, and hence the transmission of the signal by the antenna is an actual cause of the signal received by the radio.

Halpern and Hitchcock (2011) emphasize that what we take to be normal or default conditions is to a significant extent a pragmatic and context-dependent question. They wish to contrast this with the structural equations, which they take to be objective relative to a given choice of variables. Yet it is unclear that such a clean distinction between objective and context-dependent factors is possible. As they themselves emphasize, choosing variables and their ranges is a context-dependent matter. But even once we have fixed on a set of variables and their ranges, nature does not determine a unique causal structural model and still leaves some leeway to the modeler. Thus, in the case of continuous variables, we might choose a particular structural equation partly based on the simplicity of the functional relationship posited; in the case of probabilistic models, a modeler has some freedom in deciding whether to include edges (that is, direct dependencies) between variables in the model that would represent extremely weak probabilistic dependencies.

Conversely, even though some of the factors entering into assumptions about normality or typicality are strongly context sensitive, others are much more robust and may even be underwritten by the physical laws used to construct a model. As Tim Maudlin has argued (Maudlin 2007), a distinction between default behavior and deviations from it is inherent in Newton's laws, which distinguish inertial from non-inertial motions. But even in the context of Newtonian laws, the distinction is not fully context-independent and objective, since it is not all too difficult to imagine contexts in which we would be strongly inclined to view the default motion of an object to be non-inertial motion. For example, the default motion of a satellite orbiting Earth in many contexts might be taken to be its orbital non-inertial motion, and in such a context a rocket propulsion that has the effect that the satellite moves inertially would be taken to be a cause of inertial motion – that is, of a type of motion that in other contexts would be thought to be uncaused.

Introducing the notion of a body's Newtonian default behavior also allows us to understand why one might have different intuitions concerning the question whether the past state of an object in inertial motion is a cause of its present state. One the one hand, the past state is part of the causal structure into which the present state is embedded; on the other hand, the past motion is not an *actual cause*, if the body's inertial motion is assumed to be its default behavior.

Adding a "normality theory" to an account of causation serves to highlight what I take to be the Janus-faced character of causal judgments. On the one hand, causal claims aim to capture something objective and context-independent: they aim to capture facts about the causal web into which a phenomenon is embedded – or at least about the causal web as represented by us. On the other hand, causal claims feature in explanations, and the concept of explanation is inherently pragmatic. What an appropriate explanation of an event is, is highly context-sensitive. Thus, when people disagree on causal claims, the disagreement might be about the explanatory relevance or salience of a given claim rather than about the underlying causal structure. A normality theory provides one way in which explanatorily salient causal factors can, depending on context, be distinguished from less salient ones. In our example, a zero incoming field is taken to be the default state of the free field and is not explanatorily relevant to the radio's reception. Similarly, the presence of oxygen when a match is struck against the box can be seen as part of the normal or default conditions and hence not explanatorily salient in a certain context to the appearance of the flame, even though the presence of oxygen is part of the causal web.

5. Conclusion

In this chapter I examined several arguments for the claim that causal notions are incompatible, or at least sit ill, with modeling in our established theories of physics. The arguments I examined appealed to the notions of coarse-graining and background condition. As we have seen, none of the arguments is convincing. I also argued that the fact that considerations of normality or of default behavior play an important role in causal judgments is yet another aspect of the human face of causation that does not point to an incompatibility of causal notions with physics. In the next chapter I will examine another set of anti-causal arguments that appeal to a broadly interventionist conception of causation and will argue that these arguments are similarly unsuccessful. None of this shows, however, that causal notions do in fact play a legitimate role in physics. Arguments for the positive thesis will have to wait until Chapter 5.

CHAPTER 4

Causation and intervention

1. Introduction

Many neo-Russellians share the view that connections between causal concepts and those of manipulation, intervention, or agency lie at the very core of the notion of cause; indeed, the idea that these two clusters of concepts are closely linked has great intuitive appeal. Causal relations provide us with the means for manipulating or controlling a system and, as Cartwright has argued, the distinction between cause-effect relationships and mere correlations is needed to ground the distinction between effective and ineffective strategies: an effective strategy proceeds by intervening into a cause of the desired outcome and not merely into an event that is merely correlated with the outcome. Only causal relations but not mere correlations are exploitable by us in order to bring about a certain outcome. By the same token, we can test for causal relationships and distinguish them from mere correlations by selectively intervening into a system.

The notion of intervention is obviously at the core of agency accounts of causation, which take the causal asymmetry to be reducible to a more fundamental asymmetry characterizing our perspective as agents (Menzies and Price 1993; Price and Weslake 2009). The same holds, however, for Albert and Loewer's "thermodynamic" account of causation, which takes the notion of a *causal handle* (Albert 2000) or of *decision counterfactuals* (Loewer 2007) to be central to our notion of cause (see also Healey 1983); it is also true of the non-reductive manipulability or interventionist accounts of Pearl (2000, 2009) and Woodward (2003, 2007).

Some neo-Russellians have argued that the tight connection between causal and interventionist notions provides additional reasons for doubting that causal notions can play a legitimate role in fundamental physics. The most extensive discussion of this worry occurs in Woodward (2007), and I will largely focus on Woodward's arguments here even though other "neo-Russellians," such as Hitchcock or Pearl, have raised similar objections.

The central idea of Woodward's account of causation is that the results of interventions into a system are a guide to the causal structure exhibited by the system. By intervening into a system we can "wiggle" the value of a variable independently of that variable's other causes and then trace how the wiggles percolate through the system, allowing us to acquire information about the system's causal structure. Woodward defines the notion of *total cause* in terms of the notions of intervention as follows:

> (C) X is a total cause of Y if and only if under an intervention that changes the value of X (with no other intervention occurring) there is an associated change in the value of Y. (Woodward 2007, 73)

Yet Woodward's account is not a reductive account, since the notion of intervention for him (unlike for agency accounts of causation, such as the one defended in Menzies and Price [1993]) is itself a causal notion and the notion of an intervention variable I is defined as follows:

I is an intervention variable on X with respect to Y, if and only if I meets the following conditions:

(IV)
(1) I causes X.
(2) I acts as a switch for all the other variables that cause X. That is, certain values of I are such that when I attains those values, X ceases to depend upon the value of other variables that cause X and instead only depends on the value taken by I.
(3) Any direct path from I to Y goes through X. That is, I does not directly cause Y and is not a cause of any causes of Y that are distinct from X except, of course, for those causes of Y, if any, that are built into the I-X-Y connection itself; that is, except for (a) any causes of Y that are effects of X (i.e., variables that are causally between X and Y) and (b) any causes of y that are between I and X and have no effect on Y independently of X.
(4) I is independent of any variable Z that causes Y and is on a directed path from I to Y that does not go through X. (Woodward 2007, 75)

As a reductive analysis of the notion of total cause in terms of that of intervention, the account would be circular. Instead the aim of the account is to elucidate the concept of cause by making precise various conceptual connections between that concept and several related concepts, such as that of intervention or of counterfactual dependence.

In an earlier paper, Woodward and Daniel Hausman proposed a definition of the notion of an arrow-breaking intervention Z_i on X_i with respect to the variable set V (Hausman and Woodward 2004):

> (HW) Z_i is an arrow-breaking intervention on X_i with respect to the variable set V if and only if
>
> (1) Z_i satisfies the error-variable idealization.
>
> (2) If X_i is deterministically caused, then for some range of values of Z_i, z_i^*, if $Z_i = z_i^* \in z_i^*$ [*sic*], then $X_i = x_i^*$, regardless of the values of any of the X_s, U_s, or Z_s. . . . We shall say that z_i^* consists of the 'on' values of Z_i. For other values of Z_i, X_i . . . is a function of [the parents of X_i] and other omitted causes, the effect of which is summarized by U_i (Hausman and Woodward 2004, 149–50).

Z_i satisfies the error-variable idealization exactly if:

(i) Z_i causes X_i.

(ii) Z_i is not caused by any of the X_s, U_s, or other Z_s.

(iii) Z_i does not cause any of the U_s (nor, as (ii) implies, any of the other Z_s) and has no causes in common with any of the U_s or other Z_s.

(iv) For all X_j for $j \neq i$, if Z_i or any cause of Z_i causes X_j, then it does so only via a path passing through Z_i and X_i first (Hausman and Woodward 2004, 150).

Woodward and Hausman informally explain the action of an intervention as follows: "When Z_i has one of its 'on' values, it fixes the value of X_i and turns off the connection between X_i and its other causes" (*ibid.*).

There are three distinct reasons why one might think that models in physics are incompatible with interventions satisfying (IV) or (HW): First, one may question whether and to what extent it is possible to apply an "arrow-breaking" conception of interventions to models in physics, as Woodward argues: "the fact remains that in many physics contexts there may be no physically realistic operation corresponding to placing some variable of interest entirely under the control of an intervention variable, and breaking all other causal arrows directed into it" (Woodward 2007, 94–5). I will discuss this worry in Section 2. Second, Pearl, Woodward, Hitchcock, and Hausman have argued that for the notion of intervention to be applicable to a system, the system has to have an "outside," and hence interventionist notions cannot legitimately be applied within the context of putatively universal physical theories that have models of the universe as a whole among their sets of models. I will address this concern in Section 3. Finally, (IV) requires that we be able to distinguish between distinct causal paths in a system, and one might worry that this is impossible in systems modeled with the help of physical laws. This objection will be the focus of Section 4.

2. Arrow breaking

The first reason that I want to examine for doubting the applicability of an interventionist notion of causation to physics is connected to the claim that an intervention into a variable X has to remove any causal dependence of X on any variables other than the intervention variable I – that interventions apparently have to be "arrow-breaking." Woodward argues that this condition can be satisfied in the case of causal models in the special sciences precisely because causal relations in the special sciences have only a limited range of invariance: "These assumptions about the possibility of turning off or breaking certain causal influences in order to isolate and investigate others go hand in hand with the fact that causal generalizations on which common sense and the special sciences focus have only limited ranges of invariance and stability" (Woodward 2007, 94). Interventions, it seems, correspond to a change in the system that take the system outside of the range of invariance of the causal generalizations characterizing the system. Since the dynamical laws of the established theories of physics are taken to have unlimited ranges of invariance, (IV)- or (HW)-style interventions are impossible in physical systems.

As a first reply I want to reiterate a point I made in the previous chapter: to the extent that the possibility of intervention depends on taking a system outside of the range of invariance of its governing laws, the resulting anti-causal argument applies only to theories of truly unlimited range of invariance. Hence, the argument presupposes a certain conception of physics according to which there are such laws and then only applies to the most fundamental theories of physics, such as perhaps a putatively ultimate theory of quantum gravity, which might be thought to have truly unlimited range of invariance. If, as I urged in Chapter 2, we thought that each law, no matter how "fundamental," had its proper domain of application – the range of phenomena that we model with its help – then Woodward's putative distinction between physics and the special sciences collapses. On this non-foundationalist view, the proper domain of application of quantum-field theory, for example, is restricted to the production and annihilation channels of elementary particles, examined in high-energy particle accelerators. As we have seen in the last chapter, the physical processes that interact with the production and annihilation of elementary particles are not, and in fact cannot be, modeled quantum field theoretically. Instead physicists use resources from theories such as classical electrodynamics, fluid dynamics, and solid-state physics to model the causal structure within which the quantum-field theoretic interaction is embedded. A fortiori, various interventions into the accelerator or the detector systems can be modeled using these theories without having to worry about how these interventions might be modeled within an imagined enormously complex quantum field theoretic representation.

Second, it is far from clear that arrow-breaking interventions can occur only by taking a system outside of the range of invariance of the laws governing the system. Woodward suggests that disrupting a causal relationship between two variables *A* and *B* goes hand in hand with causal relations having only a limited range of invariance in the special sciences. But it is important that we distinguish between causal relationships that hold within a particular type of causal structure relative to certain boundary conditions and more general laws, which may underwrite these relationships. Often the causal relationships between variables in a model will be derived generalizations that hold only subject to certain boundary or background assumptions. Thus, it is in principle possible to break such relationships *either* through changes in the boundary conditions *or* by going outside the range of invariance of the governing laws. Imagine, for example, a complex mechanical system consisting of pulleys and ropes. Certain functional dependencies will hold between the forces exerted on the ropes and the motion of the load attached to the pulleys. These dependencies are derived from Newton's laws and constraints or boundary conditions characterizing the system. Putatively causal relationships may be broken by cutting one of the ropes – that is, by changing the system's boundary conditions – while leaving the laws of mechanics intact.

One might reply that not every putative intervention into a physical system can be characterized as a change in boundary constraints in this manner, and that there is a large class of putative interventions that cannot be characterized as (IV) or (HW) interventions because of the basic character of physical laws. Consider Newton's laws of motion. Promising candidates for causes of a change in an object's state of motion are the forces $F_1 + F_2 + \cdots + F_n$ acting on the object:

$$d\mathbf{p}/dt = \mathbf{F}_{Total} = \mathbf{F}_1 + \mathbf{F}_2 + \cdots + \mathbf{F}_n.$$

A paradigmatic Newtonian intervention into an object's motion would arguably have to be represented by an additional force F_i, which would result in a change of the total force acting on the object from $\mathbf{F}_{Total}$ to $\mathbf{F}_{Total} + \mathbf{F}_i$:

$$d\mathbf{p}i/dt = \mathbf{F}_{Total} + \mathbf{F}_i.$$

But if interventions would have to be modeled by the introduction of an additional force, then there could be no arrow-breaking interventions in physics, since introducing an additional force neither erases any of the other forces nor shields the object against other forces – in fact, in the case of at least one fundamental physical force, the gravitational force, such a shielding is impossible. $\mathbf{F}_i$ represents an additional force but does not break any causal arrows represented by the other forces. This suggests the following instance of the anti-causal argument schema that we already encountered in the last chapter:

2.1   Causal notions can only legitimately be applied in contexts that can be represented as permitting arrow-breaking interventions.

2.2   Arrow-breaking interventions cannot be modeled with the help of our established theories of physics.

2.3   Therefore, causal notions play no legitimate role in physics.

In what follows I want to develop a series of different replies to the foregoing argument. First, it is important to distinguish the switch condition (IV.2) or condition (HW-2) from Woodward and Hausman's intuitive characterization of these conditions as "turning off" the connection between the variable intervened into and other variables in the model. As we will see, the switch condition (IV.2) can be satisfied even for interventions that are not "arrow-breaking" in an intuitive sense and do not disrupt or turn off the causal connections between the variable intervened into and its causal parents. Thus, even if it were true that representations in physics do not allow for arrow-breaking interventions, this does not imply that Woodward's notion of an intervention does not apply in physics. Second, the desideratum that there be a conceptual link between the notions of causation and manipulation can be satisfied even if interventions into a system do not satisfy the switch condition. Thus, not only is the condition of arrow-breaking too strong; the switch condition is so as well. And third, I will argue that, initial impressions to the contrary, there are systems and interactions even in the domain of our established theories of physics that are legitimately modeled as involving arrow-breaking interventions. Later I will also consider a worry that arises from the "error-variable-idealization" conditions (ii) to (iv) that require interventions to come "from outer space," as Hausman and Woodward say, and require the possibility of isolating distinct causal paths.

Figure 4.1 A causal chain

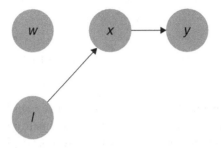

Figure 4.2 An arrow-breaking intervention

To begin, let us consider a simple example of an intervention. Take a model of the trajectories of a small number of billiard balls on a pool table. Let the variable X be the position of ball x at time t, which according to the model depends on x's prior trajectory and, hence, on any collisions of x with other balls. Let an intervention represented by the intervention variable I consist of my reaching into the table to place the ball x at some specific location determined by me, independently of its present state of motion. My action acts as a switch in Woodward's sense, in that it renders the position of the pool ball counterfactually independent of its prior state of motion and of its motion's causes, and therefore constitutes a genuine intervention, according to (IV) or (HW).

There are two distinct ways, however, in which we can represent this intervention in a causal model. First, we can represent my reaching in as a genuinely 'arrow-breaking' intervention, which erases all causal arrows W into X except for one from the intervention variable I. Consider the causal chain in Figure 4.1. An arrow-breaking intervention, which breaks the causal dependence of X on W, is represented in Figure 4.2. Alternatively, we can represent the intervention in terms of two distinct variables: a control variable I_C, which like Woodward's intervention variable controls the value to which we want to set X; and a second variable I_F, whose values depend both on the value of the control variable and on the prior state of the ball x (see Figure 4.3). As a matter of fact, if my aim is to place the ball at a specific location, my intervention has to take the ball's prior state of motion

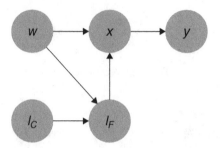

Figure 4.3 A "feedback" intervention

into account. I have to reach into the table at the right location and exert just the right force on the ball to change its momentum in just the right way. I could represent this aspect of the intervention in a causal model by including additional causal arrows from the causal parents of X into a "feedback" intervention variable I_F. These additional arrows represent that I_F in a sense "measures" the values of X's causes and adjusts its own value accordingly, so as to ensure that X takes on the value set by I_C independently of the values of any of X's causes. This second representation, which has the structure of a feedback control system, also satisfies Woodward's switch condition in that it renders the value of X *counterfactually independent* of its other causes, even though it does not "turn off" or "break" the causal links between X and its causal parents.

In fact, the two kinds of representations are closely related: An arrow-breaking representation of the intervention can be understood as a reduced representation of the feedback system, which will adequately represent the effect of the intervention variable I on the variable intervened into X exactly if the feedback representation adequately represents the effect of the control variable I_C on X. The equivalence between the two representations follows directly from the functional equations of the respective models, since X depends on I_F in such a way that any dependence of X on its other causal parents is exactly canceled and X is set to whatever value is dictated by I_F's other parent – the control variable I_C. Prior to the interventions, the value of X is given by a structural equation of the form $X = f(PX)$, where PX are the causal parents of X. If we model the intervention as feedback intervention, then the resulting structural equation for X is $X = f(PX) + f(I_F)$, where $f(I_F) = -f(PX) + f(I_C)$. That is, after the intervention, X is rendered counterfactually independent of the values of its causal parents, even though no causal "arrows" are in fact broken. Nevertheless, we can then choose to represent the intervention in terms of an arrow-breaking single intervention variable, with the corresponding equation $X = f(I)$. That is, in circumstances where the feedback model provides an appropriate representation, the arrow-breaking model can equally be used as a simplified representation that agrees with the arrow-breaking model on all counterfactual dependencies among all variables that are part of the latter model.

I want to draw two lessons from the discussion so far. First, there are causal structures that satisfy the switch condition but are not intuitively arrow-breaking. Similarly, no causal arrows need to be literally broken in interventions that are (HW)-arrow-breaking. Hence, one can accept both that the switch condition is a necessary condition on interventions into causal structures and that models of physical systems do not allow for arrow-breaking interventions and yet resist the conclusion that an interventionist notion of causation satisfying (IV) is inapplicable in physics. It is not necessary that we have the "possibility of turning off or breaking certain causal influences," as Woodward says, or of taking a system outside of the range of invariance of its laws in order to intervene into a system in a way that renders a variable counterfactually independent of its causal parent. Thus, in the Newtonian scheme introduced earlier, if an intervention proceeded by adding a force $-\mathbf{F}_{Total} + \mathbf{F}_i$ to the total force prior to the intervention, then the switch condition would be satisfied even though no causal arrows are broken or erased. Second, the fact that arrow-breaking models can provide reduced representations of non-arrow-breaking models suggests that it does not follow from the fact that the underlying physics might not allow genuine arrow-breaking that arrow-breaking models are inapplicable to physical systems. Even though an arrow-breaking model leaves out certain details of the joint system of intervention plus system intervened into, it may nevertheless provide us with an adequate representation of the system modeled.

Although modeling an intervention as adding a feedback system might appear to be more realistic from the perspective of the underlying physics than an arrow-breaking model, feedback models face what some take to be a devastating problem. A causal model in which the causal dependencies along different routes exactly cancel, and hence a model in which the causal dependence of X on (some of) its causes is not reflected in a probabilistic or counterfactual dependence of X on the causes in question, does not satisfy a condition that is frequently imposed in the causal modeling literature as necessary condition on any causal structure: the condition that any cause of a variable X is probabilistically correlated with X – or, in the terminology introduced by Patrick Suppes (1970), that every cause of X is a *prima facie cause* of X. Judea Pearl calls such a condition *stability* (see Pearl 2009, 48), and Spirtes, Glymour, and Scheines call it *faithfulness* (Spirtes et al. 2000, 35; see also Hausman and Woodward 2004). But it is unclear whether the correct conclusion to draw from the violation of stability is that one cannot model switches as feedback systems, since this would violate stability, or whether we should conclude instead that the stability or faithfulness condition cannot be a general constraint on all causal models.

Spirtes, Glymour, and Scheines argue that parametrizations that violate faithfulness have probability zero in a linear causal model (2000, theorem 3.2, p. 42). The argument proceeds in two steps. First, they show that any parametrization violating faithfulness has Lebesgue measure zero. The second step is to that any subset of parameter space that has Lebesgue measure zero also has probability zero. Applied to our case, the argument says that the subset of parameter space, for which $f(I_F) = -f(PX) + f(I_C)$ and the dependence of X on its causal parents is exactly canceled by the dependence of X on I_F, has Lebesgue measure zero and, hence, has probability zero.

Against the argument in Spirtes et al. (2000), Cartwright (2001) and Kevin Hoover (2001) have claimed that there are many actual systems in which stability is violated and, hence, that we should not expect such systems to be uncommon. Since there exist systems that either have been explicitly designed or have evolved to contain feedback mechanisms, for which a representation in terms of a causal structure that violates stability would be a natural choice, the inference from the claim that a set of parametrizations has Lebesgue measure zero to the conclusion that it has probability zero ought to be resisted. As an example from biology, consider body temperature, which (within a certain range of invariance and to a certain degree of approximation) is counterfactually independent of ambient temperature, precisely because the human body responds to changes in ambient temperature through different mechanisms along different causal routes, whose function is to maintain a constant temperature. The external heat transfer mechanisms are radiation, conduction, convection, and evaporation of perspiration. These mechanisms do not merely operate passively, but are exploited by neural feedback mechanisms. For example, sweating begins almost precisely at 37°C and increases as the skin temperature rises above this value; there are also a variety of mechanisms directed at conserving body temperature, such as shivering to increase heat production in the muscles or vasoconstriction to decrease the flow of heat to the skin.

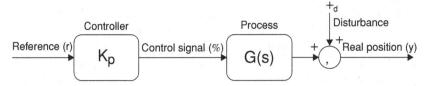

Figure 4.4 Representation of the control system at the LHC (from Steinhagen 2007)

The LHC provides us with an example from physics that has been designed to contain sophisticated feedback systems that are best represented in terms of a stability-violating feedback model of interventions (Steinhagen 2007). There are various sources of disturbance of the proton beam, both environmental and within the accelerator, which can lead to losses of particles inside the machine. The disturbances are counteracted through feedback systems designed to keep the orbit stable by controlling key beam parameters, such as its orbit, its energy, and its tune (defined as the number of transverse oscillations a particle describes per turn). The size of the LHC feedback system is impressive: the beam control system consists of more than 1,000 beam position monitors and more than 1,000 orbit corrector magnets, whose fields are adjusted according to calculations of a central global feedback controller. The control system is naturally represented as causally intervening into the proton beam to keep it stable and to prevent damages to the accelerator: The beam is one physical system with a certain time evolution into which a second system – the beam control system – intervenes. Indeed, Steinhagen explicitly represents the control system causally in terms of box-and-arrow-diagram in which "the signal flow and causality are indicated through arrows" (40) (see Figure 4.4).

Even though the effect of the control system (ideally) is to render the variables characterizing the orbit independent of the size of prior perturbations, as Woodward's condition (IV.2) on intervention requires, the intervention does not proceed by breaking the "causal arrows" going into the beam variables but by being extremely sensitive in its responses to measurements of the prior state of the system intervened into.

In defense of a stability condition, one might argue that exact cancellations along different routes will nevertheless be extremely rare.[1] Body temperature will not stay *exactly* constant, as ambient temperature changes and the controlled beam parameters in the LHC are not *completely*

---

[1] See also the discussion in Cartwright (2001, 253–4).

independent of the prior particle orbits. Thus, one might want to insist that the stability or faithfulness condition can be saved and that feedback systems of the kind I posited do not offer a genuine alternative model of interventions to an arrow-breaking model, since they do not in fact satisfy the "switch" condition. Indeed, Daniel Steel (2006) argues that if the variables whose values represent the parameter values in a causal model are jointly continuous and values of the variable are allowed to vary independently, then strict violations of stability will be extremely rare, even if the distribution of the values is tightly focused on a Lebesgue measure zero subset of the parameter space.

Yet here it is once again important to distinguish carefully between causal models and the real-world systems they are intended to represent. If a feedback control system works so well that the quantity controlled remains constant within the bounds of experimental error, or if the variations in the quantity are small enough to be ignored given a certain context and certain interests, then it seems to be appropriate to represent the quantity at issue in a causal model by a variable whose values are strictly independent of the variable's other causes, given the operation of the feedback system. If we nevertheless were to insist that the causal model has to satisfy stability, then this would require that we introduce a more fine-grained dependence of the variable on its causes than is warranted either by our measurement precision or by the context in which the model is used.

Moreover, appealing to fine-grained dependencies in order to rescue stability leads to an odd dialectical situation as far as our main issue is concerned – the place of causal notions in physics. For the objection seems to imply that the stability condition can be expected to hold in general only once we move to especially fine-grained models of physical systems and not for the much more coarse-grained and arguably more highly idealized models of the higher sciences. To the extent, then, that *stability* is a necessary condition on causal models, fundamental physics seems to be much more hospitable to causality than the higher sciences, where, as our models get ever more coarse-grained and idealized, we would expect failures of stability to become more and more common. By contrast, Woodward's *switch condition* seems to be satisfied only in the higher sciences, if the objection is to be believed.

Pearl also argues for the stability condition by maintaining that it is a distinctive feature of causal relations between different pairs of variables that they can vary independently of one another – that different causal links are "autonomous" from one another – and, hence, that exact cancellations along different causal routes are unlikely and "will rarely occur under natural conditions" (Pearl 2009, 63). But whatever the force of Pearl's argument may be, we cannot help ourselves to it in the present context, because what is at issue currently are precisely *non-natural* (merely imagined our actual) interventions into physical systems, that are carefully set up just in the right way to allow us to probe a system's causal structure.

This concludes my first reply: I have argued that a condition that demands that interventions be genuinely arrow-breaking is too strong: the switch condition and Hausman and Woodward's condition can be satisfied without arrow-breaking. Hence, the alleged impossibility of introducing arrow-breaking interventions into models of physical systems does not imply that the intervention conditions (IV) or (HW) cannot be satisfied. My second reply is that the switch condition is too strong as well. We can account for the intuitive link between interventions and causal relationships also if we allow for a weaker notion of intervention that does not satisfy the condition. Consider the following variant of the billiard-ball example given earlier. Instead of reaching in with my hand, I give the ball an extra nudge with the cue. My action will change the position of ball x from what it would have been without my nudging it, but it does not render the ball's position counterfactually independent of its other causes, such as the earlier motion of other pool balls: the final position of the ball depends *both* on the nudge it receives from the cue *and* on the ball's prior state of motion. Even though my action does not satisfy the switch condition, it seems to constitute what intuitively is an intervention into the state of the balls. Earlier I suggested that, more generally, adding another force or changing the value of a component force acting on an object appears to be a paradigmatic way of intervening into the state of the object. Rather than maintaining that these putative "intervenings" are not genuine interventions and concluding from this that physics is inhospitable to causal notions, one could instead argue that Hausman and Woodward's notions of intervention are too restrictive.

Hausman and Woodward motivate their condition (HW) by arguing that the condition, together with a modularity assumption, implies the causal Markov condition. The modularity assumption states that an intervention with respect to a variable X_i does not change the probability distribution of any other variable X_j that is not an effect of X_i. According to the causal Markov condition, if X_i does not cause X_j, then X_i and X_j will be probabilistically independent conditional on the causal parents pa_i of X_i. But the proof of the causal Markov condition given by Hausman and Woodward does not require (HW) as a premise but only invokes the weaker assumption that an intervention variable "sets" the distribution of the values of X_i conditional on the causal parents pa_i of X_i: all that is required is that "conditional on pa_i, the [exogenous variables] satisfy the definition of an intervention" (Hausman and Woodward 2004, 153) and not also that, given an intervention I_i, X_i is rendered counterfactually or probabilistically independent of its causal parents.

Thus, we can replace condition (IV.2) with a weaker condition requiring of an intervention only that it be able to *change* the value of X from what it would have been otherwise, and not that it do so by breaking all other causal arrows leading into X:

> (IV 2*) I acts as an intervention on X. That is, certain values of I are such that when I attains those values, the value of X is a function of I together with the parents of X, PX (and $f(PX, I) \neq f(PX)$).

On this weaker notion of intervention, an intervention "sets" the value of the variable X intervened into, conditional on the set of other causal parents pa of X. From this (and modularity), the causal Markov condition follows immediately: Modularity implies that X_j is independent of the intervention I, conditional on the causal parents pa_i of X_i, for all $X_j \neq X_i$ that are not effects of X_i. That is, $P(X_j \mid I \& pa_i) = P(X_j \mid pa_i)$. For deterministic theories, X_i is a function of $I \& pa_i$, and hence: $P(X_j \mid X_i \& pa_i) = P(X_j \mid pa_i)$, which is a statement of the causal Markov condition. In the case of indeterministic theories, this last step involves as an additional substantive assumption that no spontaneous correlations exist, as Hausman and Woodward explain.

This weaker notion of intervention is discussed in the literature, where it is sometimes referred to either as "soft intervention" (and is contrasted with "hard" interventions) or as "parametric interventions" (and is contrasted with "structural" interventions) (see, e.g., Eberhardt and Scheines, 2007). We can once again turn to the LHC for an example, this time of a soft intervention. Slow particle losses in the machine are minimized with the help of a "beam cleaning system" consisting of a series of collimators that absorb slow particles. Depending on the precise trajectory of a slow particle, the primary collimator will either absorb the particle or scatter it in a manner such that the particle will be absorbed by a secondary (or even tertiary) collimator. That is, the beam cleaning system intervenes into the slow particles of the beam but in ways that do not render the motion of a particle independent of its prior state: the trajectory of a particle depends both on its prior state of motion and on the action of the collimator. That is, the collimators do not produce hard interventions, acting as a switch, but soft interventions that alter the trajectory of slow particles.[2]

If we believe, with Woodward, that "causal notions and patterns of reasoning seem less appropriate when applied to physical systems" since "in many physics contexts there may be no physically realistic operation corresponding to placing some variable of interest entirely under the control of an intervention variable," we are also committed to the claim that causal notions are only fully appropriate in contexts where hard interventions are possible. But this last claim is problematic for at least the following two reasons. First, as Frederick Eberhardt and Scheines show, soft or parametric interventions can in certain contexts be more efficient in discovering the causal structure of a system. This is the case, if multiple interventions are allowed per experiment. Intuitively, the drawback of hard interventions is that they destroy the causal structure of the system intervened into. Second, arguably there are other sciences, besides physics, that do not allow for hard interventions. Thus, John Campbell has convincingly argued that there are no or only few realistic operations that would correspond to hard interventions on mental states (Campbell 2006). The typical intervention here is not one in which an agent's rational autonomy is suspended by giving complete control over the agent's mental states to the intervening subject. Rather, an intervention makes a difference to an agent's decisions or intentions, given the agent's other mental states, by providing additional reasons or motivations. Campbell also points to a paradigmatic example of an intervention in physiology that is not a hard intervention: the administration of a drug that intervenes in the organism without, blocking any endogenous production of the drug in the organism. Thus, to the extent that physical systems do not permit hard interventions, this feature does not point to a genuine contrast between physics, on the one hand, and the special sciences, on the other. Here, too, there is no contrast between physics and the special sciences of the kind posited by the neo-Russellians.

A third reply to the argument from arrow-breaking is that it is far from clear that models that *do* posit genuine arrow-breaking cannot be used to

---

[2] In fact, Hausman and Woodward themselves suggest that such an alternative conception of intervention is possible (see also Hausman 1998, Chapter 5): "One can also conceive of an intervention as the addition of an intervention variable into a graph with an arrow into the variable that is manipulated that leaves all the arrow in the graph intact. An intervention might, for example, merely add to the value of a variable without cancelling the effect of its causes" (Hausman and Woodward 1999, 542, fn. 14).

represent systems in physics. To reiterate a point I made in the previous chapter, a causal model of a physical system need not, and arguably cannot, be complete and represent all quantities and interactions that are part of the system – any model we construct, no matter at which level of "grain," will abstract and idealize in some way or other. If we keep in mind that in looking for operation that represent arrow-breaking interventions, what is at issue is whether there are physically realistic operations that *in a given context are appropriately represented* as involving an intervention variable that breaks all other causal arrows directed into the variable at issue, then it is not so difficult to give examples of such circumstances. Thus, an arrow-breaking model may be appropriate in cases in which the intervention under consideration effectively swamps all other influences on the variable intervened into and therefore allows us to ignore any in-principle dependence of the variable's values on its other causes. Again we can illustrate this with an example from the LHC. Certain failures in the LHC can trigger a "beam dump" – that is, the extraction of the proton beams from the accelerator into a large absorber. Even if we thought that events in any of the detectors at one time *in principle* depend on everything in the events' backward lightcones and, hence, also on the state of the accelerator at earlier times even when no proton beams are present in the accelerator, we do not need to explicitly represent this in-principle dependence in our model of the accelerator and detectors. Rather, given how small any influence of the state of the beam dumps is on the detectors, the triggering mechanism arguably ought to be represented as an intervention that breaks the causal arrows from the goings-on in the accelerator to the states of the different particle detectors.

To sum up this section, I have responded to Woodward's worry that the condition that interventions are arrow-breaking will not be satisfied in models constructed with the help of fundamental physical laws. My response was threefold. First, I argued that there can be non-arrow-breaking interventions that nevertheless act as switches in Woodward's sense and render a variable intervened into counterfactually independent of its other causes. Second, I argued that the notion of hard intervention, or interventions that act as switches, is too narrow and that a notion of soft intervention that merely assumes that interventions change the value of a variable intervened into without breaking all other causal arrows into that variable is sufficient to underwrite causal structures. Finally, I argued that once we grant that all scientific representations idealize and distort in some ways or others, there is no in-principle obstacle to representing physical interactions in terms of genuinely arrow-breaking interventions, even in cases where we believe that an in-principle dependence on the severed causal links persists.

3. Large and small worlds

I now want to turn to a cluster of worries concerning the applicability of the notion of intervention that are associated with putative spatiotemporal properties of the representations we construct with the help of our physical theories. First, it seems to follow directly from Field's claim that physics provides us with models of a spatial region R at time t that require as input an entire cross section C of the backward lightcone of R that the notion of intervention cannot be applied to models in physics. For if physical models consist of entire backward lightcones (up to the cross section C on which the initial data are specified), then interventions into the model are physically impossible. Any putative intervention I_{out} originating outside of the backward lightcone could not reach R at t without involving faster-than-light propagation, while any variable I_{in} representing an event inside of the backward lightcone of R would already have to be included in the model as one of R's putative causes, and therefore could not represent an external intervention into the system. Thus, external interventions into backward-lightcone models are physically impossible.

In light of our discussion in the last chapter, my reply to this worry will be obvious: many, or even most, models constructed even with the help of our most fundamental theories do not involve specification of the state of the full backward lightcone of the spacetime region we are modeling, but also include spatial boundary conditions. Thus, even if we accept Field's and Woodward's claim that the representation of a physical system does not involve any spatial or temporal gaps, this does not imply that there is no room for interventions into physical systems: interventions can occur across the spatial boundaries of the system and can be represented as a localized change in the boundary conditions.

This reply presupposes, however, that the system in question has an outside from which we can intervene, yet at least some of our mature physical theories also provide us with models representing the universe as a whole and such models, it seems, leave no room for an external vantage point or an "outside" from which to intervene into the model.

That it is difficult or even impossible to locate causal relations within theories that have potentially universal scope is a view shared by many writers who have sympathies for interventionist accounts of causation. Thus, Pearl maintains that "if you wish to include the whole universe in the model, causality disappears because interventions disappear" (2000, 350); Hitchcock claims that "how to find or even understand causation from within the framework of a *universal* theory is one of the very deep problems of philosophy" (Hitchcock 2007, 53); and even though Woodward is more tentative in his skepticism, he also believes that causal ascriptions become "increasingly strained when candidate causes expand to include the state of the entire universe" (Woodward 2007, 93). The worry, as Hitchcock states it, is both epistemological and conceptual: what reasons can we have for attributing causal relations to a model of the universe as a whole? And how do we even understand putatively causal relations that cannot, even in principle, be exploited through an external intervention?

I want to address the second worry first. Let us grant that it would be meaningless or empty to attribute a causal structure to a system to which the notion of an intervention could not even in principle be applied. As Woodward has emphasized, however, it is not necessary to assume that an intervention into a system is physically possible, but only that an intervention is logically or conceptually possible, in order for it to be meaningful to attribute a causal structure to a system. "What is crucial," he writes, is that counterfactuals describing what would happen to Y (or in the indeterministic case the probability distribution of Y) under an intervention on X "make sense" and "have determinate truth values," and not that human beings "are able to carry out the interventions in question" (Woodward 2007, 91).

Alexander Reutlinger (2013) argues that this shows that an anti-causal argument appealing to the "large worlds" fails, since it is logically or conceptually possible for the universe to contain more objects than it actually contains. Thus, a counterfactual intervention into a variable that is part of a global model (representing the entire content of the actual universe) can be thought of as involving an additional variable representing some object posited counterfactually to exist in addition to the universe's actual content. Yet this strategy of counterfactually adding another intervention variable to an existing global model is not available in the case of a global model defined over all of spacetime. In this case there simply is no extra spatiotemporal location to which an object could be added and from which the intervention could occur.

There is a more promising strategy available, however, and that is to invoke an alternative notion of intervention that does not require positing an additional intervention variable outside the system intervened into and which can unproblematically be applied even to a global model of the universe as a whole. Ironically, Pearl himself formally introduces such a notion of intervention (Pearl 2000, 69f). As we have already seen, a *causal model*, for Pearl, consists of a directed acyclic graph over a set of variables $\{X, Y, \ldots\}$, structural equations $x_i = f_i(pa_i, u_i)$, which specify the value of each variable in terms of the value of the variable's causal parents pa_i and a random exogenous disturbance u_i, and a probability distribution $P(u_i)$ over the values u_i of the exogenous variables U_i. An *intervention* into the variable X_i, according to Pearl's account, then consists of removing the structural equation for X_i from the causal model and replacing it with an equation that fixes the value of X_i to some fixed x_i. That is, formally an intervention is represented by removing the equation $x_i = f_i(pa_i, u_i)$ from the model and replacing it with some x_i. Pearl calls such an intervention an "atomic intervention," which can be denoted by "*do* $(X_i = x_i)$" or "*do*(x_i)" (Pearl 2000, 70). Any more complex intervention that forces several variables to have fixed values can be represented in terms of a set of atomic interventions. Although one might think informally of interventions as requiring that a system have an environment from which the intervention is to be performed, this informal conception is not part of Pearl's do-calculus. Thus, contrary to Pearl's own worry, it makes perfect sense to ask, on his own account, how the values of variables would change under an intervention into a model of the universe as a whole. The $do(x)$ operation does not require that we assume that the causal structure on which it is performed be embedded into a larger environment represented in part by exogenous intervention variables.

One might worry, however, that counterfactual interventions require that we investigate nomically impossible situations or worlds. The dynamical laws of many of our mature theories are both past- and future-deterministic. That is, for any two dynamically possible worlds W_1 and W_2, W_1 and W_2 agree in their state at some time t, exactly if $W_1 = W_2$. Thus, if the value of some variable changed as a result of an intervention, then the dynamical laws entail not only that the evolution of the system in the future of t will be different from the system's actual evolution, but that its past evolution would have to have been different as well. If we insist that interventions into a causal structure only have an effect on the values of variables in the intervention's causal future, then one might worry that the dynamical laws governing the system cannot be applied in a meaningful or coherent way to the model, since the model is posited to have a past that is dynamically inconsistent with its state resulting from the intervention. Hence, as Woodward suggests, "it is unclear what would be involved in such an intervention and unclear how to assess what would happen if it were to occur" (Woodward 2007, 93), and one might worry that interventionist counterfactuals do not have determinate truth conditions. In the case of finite systems, as Woodward points out, modeling interventions with the help of intervention variables allows us to keep nomic violations "offstage" and restrict them to occur outside of the system of interest. When we intervene into a system from the outside, the system no longer is closed, and, thus, the fact that the system's future evolution is different from what it would have been otherwise is no threat to determinism. In the case of global models, this move is not available. An intervention results in a dynamically impossible model, and this, one might worry, threatens to render the whole approach incoherent.

Yet once we posit a *causal* model for the universe, the model provides us with a well-defined procedure as to how to apply the dynamical laws in evaluating the results of interventions, even if they involve a violation of the dynamical equations governing the system. Pearl's notion of an intervention, like Woodward's, involves arrow-breaking in that the structural equation for the variable X intervened into will be replaced by one that assigns X some non-actual value. As the result of an intervention, the value X no longer depends on its causal parents. Since only the arrows from the causal parents of X into X are erased, we can continue to use the original model, including information about the dynamical laws encoded in the structural equations, to determine what the *causal future* of the system would be from the intervention into X onward. In other words, we can use the state of the world at the time of the intervention as input in the dynamical laws to determine the future evolution of the world. But because the intervention changes the causal structure *leading into X* and the structural equation for X is replaced by the $do(X)$ operation, we cannot apply the dynamical laws *across* the posited intervention to evolve the state of the system at the time of the intervention backward in time. The causal structure provides us with a well-defined recipe as to where the dynamical laws can be legitimately applied and where not.

Intervention counterfactuals are time-asymmetric, just as Lewisian counterfactuals are. But, instead of attempting to ground the causal or intervention asymmetry in a counterfactual asymmetry, I am proposing that we turn Lewis's account on its head and ground the asymmetry of intervention counterfactuals in the asymmetry of the causal model within which they are evaluated. Moreover, contrary to Lewis's view, I do not believe that there is a *standard* kind of context for evaluating counterfactuals in which backtracking counterfactuals are false. Rather, we can distinguish at least two perfectly standard contexts that play a role in physics: a context in which counterfactuals are evaluated asymmetrically as causal or intervention counterfactuals and one in which counterfactuals are evaluated purely dynamically by taking the state of a system or the world at one time as input and determining both the past and future evolution of that state in accord with the dynamical laws. If the laws in question are time-reversal invariant, then changes at one time will be associated with changes in both the past and future evolution of the system.[3]

Since the notion of an intervention can be defined in a way that makes no reference to a system's environment, we can understand causation in the context of global models, even on an interventionist conception of causation. Pearl's *do(x)* operator simply does not care how "large" the world is to which it is applied. I have argued moreover that *once we posit a causal structure*, then intervention counterfactuals can have well-defined truth conditions even in global causal models. I have not yet, however, addressed Hitchcock's epistemological worry: can we ever have good reasons for postulating such causal structures within the context of a theory with universal scope? Yet this worry dissolves once we realize that even theories with potentially universal scope can be – and are, as a matter of fact – used to model finite systems embedded into an environment. For models of finite systems we can apply Pearl's notion intervention discussed earlier, or we can invoke Woodward's notion of intervention, which Pearl also introduces later in his book, and model interventions as involving the coupling of an additional exogenous variable to the causal structure. In either case, the kinds of reason we can have for positing a particular causal model for a system will be analogous to the reasons we give for systems modeled by strictly domain-restricted theories. That is, in answer to Hitchcock's worry, we "find... causation from within the context of a universal theory" in the very same way in which we find causation in theories with a limited domain: through interventions into finite systems modeled with the help of that theory.

If we assume that there are cases where we are justified in interpreting such *local* models of a theory with putatively universal scope causally, then a skeptic about causal relations in *global* models must believe that causal relations invariably disappear when a local model is embedded into a large model of the universe as a whole. But it is not exactly clear what reasons might be offered in support of this belief. We can imagine the process

---

[3] There may be yet a third kind of context in which the "statistical mechanical" counterfactual introduced by David Albert and Barry Loewer is the appropriate one to use. I will discuss Albert's and Loewer's views in Chapter 8.

of embedding local, finite models into successively larger models. Now, one might hold that at some point this process of enlarging reaches the limit of what can successfully be modeled even with our most fundamental theories. This is the view of Nancy Cartwright, who would caution that the fact that we can formally and abstractly speak of the possible worlds allowed by a theory should not mislead us into thinking that we are justified in accepting a theory as providing us with anything but local representations of particular phenomena. Cartwright argues that we do not know whether physics satisfies the principle of self-closure, that is, that there are "(in God's great Book of Nature) laws of physics that dictate everything that happens that can be reasonably taken to be in the domain of physics itself" (Cartwright 2010). Thus, Cartwright's view provides another argument against the present objection: one might simply deny that there are universal theories in physics that are properly thought of as having an in-principle universal domain of application. But let us assume, for the purposes of the present argument, that we can indeed embed local models of a putatively fundamental theory into ever larger models and eventually into possible universes – that is, models of the universe as a whole – as Woodward's and Hitchcock's arguments must assume we can.

The first question, then, is whether a causal model C associated with a local dynamical model D can be recovered as causal submodel from a larger causal model C^* associated with the larger dynamical model D^* into which D is embedded. Take a given dynamical model D with specific initial and boundary conditions. We obtain an embedding model D^* defined over a larger spacetime region if we treat the initial and boundary conditions of D themselves as dynamically determined in the larger model. If finite dynamical models of a theory suggest a particular causal interpretation, then the embedding requirement says that this interpretation can be preserved when we embed the model into a larger model. As long as we do not allow values for the variables newly added to the expanded model C^* that disrupt the causal relations within the submodel C, the causal relations included in C will be preserved in C^*. As an example, think of a causal model of a car engine. The causal relations characterizing this model will be preserved when we embed it into a causal model of the car as a whole. Clearly, there will be some settings for the variables external to the submodel representing the engine that preserve the causal relations within the submodel – among those will be the settings representing the actual situations in which we are warranted to adopt the causal model of the engine's functioning.

Now, the causal model for the engine presumably will only be invariant under some limited range of values of the variables characterizing its environment. Thus, one might worry that all I have said so far is that causal models can be recovered as submodels from larger models into which they are embedded exactly when they are thus recoverable. But local causal models of putatively universal theories are distinguished precisely by the fact that they have a very extensive and perhaps even unlimited range of invariance. In particular, if, with Field, we were to posit relativistic models in which the state of an entire cross section H of the past lightcone of an event e is taken to be the cause of the event in a *local* model M, then embedding the model defined over the section of the past lightcone of e that lies to the future of H into a model M^* defined over a larger region of spacetime cannot affect the causal structure associated with M.

The only remaining question is what happens at the final stage of our procedure of successive embeddings. Pearl, Woodward, and Hitchcock's objection entails that it is at this very last step – when we embed a model of almost the entire universe except for a very small region in the model's causal past into a model of the universe as a whole – that all causal relations disappear. But if, as the current argument grants, we can justifiably interpret each finite submodel of the model causally, and if the causal interpretation survives through an arbitrary number of stages of successively embedding a model into larger models, it is difficult to see what the argument for the claim could be that the global model could not similarly be interpreted causally.

The embedding condition I have just described is related to the modularity assumption: If the causal structure associated with the global model is modular, its substructures will survive a breaking-up of the global model into smaller local models, which are associated exogenous intervention variables, and, hence, reasons for attributing certain causal structures to local models will ipso facto be reasons for attributing a causal structure to the global model. But the modularity assumption has a different status in my argument than in Woodward's account: spelling out the notion of total cause in terms of arrow-breaking interventions requires that the modularity assumption hold for *all* causal (sub)systems. By contrast, in order to refute the claim that we can never find causation in global models of our putatively universal theories, it only needs to appear reasonable that the global causal models of *some* of our universal theories are modular. Moreover, Woodward's account requires "modularity all the way down" – that is, that we can always in principle perform an arrow-breaking intervention on any variable in the model without perturbing the rest of the causal structure. My argument only requires that a global causal model be modular up to a point: it must be possible to break up the global model into local causal models in a way that satisfies modularity, but it may be that this process may reach a limit beyond with the local models cannot be broken up further without changes in the causal structure within the resulting submodels.

I have argued that an interventionist account can coherently be applied to "large worlds" or models of the universe as a whole. Implicit in the commonsense notion of an intervention might be a conception of a system being embedded in an environment from which the intervention occurs. There is at least one formal account of intervention, however, Pearl's do-calculus, that does not make any reference to a system's environment. Woodward maintains that "it is very plausible that causal ascription becomes less natural and straightforward – increasingly strained – when candidate causes expand to include the state of the entire universe" (Woodward 2007, 93). I have already partially addressed that worry: in the case of theories that also have "small worlds" among their models, we might have good reasons for attributing causal structures even to large worlds modeled with the help of the theory. A further reply, which I will develop in more detail in Chapters 5 and 6, is that our universe exhibits a striking asymmetry: there is a sense in which initial conditions are random but final conditions are not. This asymmetry may be explained by appealing to a causal asymmetry. At the very least, however, the asymmetry between prevailing initial and final conditions can underwrite the legitimacy and usefulness of causal structures in representing the world.

4. The asymmetry of state preparation

In Section 2 I focused on the switch conditions (Woodward 2007, condition [IV.2]) in Woodward's interventionist account. One might also worry that condition (IV.3) is problematic in physics. It states, intuitively, that for an intervention on X with respect to Y, Y may not depend on I along any causal route not going through X. A similar condition is part of Hausman and Woodward's error-variable idealization. The worry is that at the level of a complete micro model of a system, the very notion of a causal path may break down, and we may not be able to distinguish and isolate from one another dependencies along different routes. At the level of physics, it seems, the values of variables affect each other in such interconnected and interwoven ways that the only possible causal representation, if one is possible at all, would be a representation according to which the complete state of a system on one initial-value surface causes the complete state at later times. But if we could no longer distinguish different causal routes connecting variables representing events on one time slice to variables representing events on another and were forced to take the relata of the causal relation to be complete time slices in a model, then one might worry that the concept of causation would have become so diluted as to have lost its usefulness.

I want to offer two replies to this challenge. First, even if causal relations in physics only relate entire time slices to one another, an interventionist causal model still allows us to capture one core aspect of the causal relation: its asymmetry. Even causal models that merely allow us to identify the past of a system as cause of its future evolution can play a useful role in physics. As I will argue in this section, even interventionist "time-slice" models enable us to capture an important temporal asymmetry associated with interventions into a physical system, the fact that interventions into a system are always future-directed. I will develop this line of argument further in Chapter 6, where I will discuss linear response theory. As we will see there, this asymmetry has important theoretical consequences for how the behavior of a system subject to an external field or force can be modeled in the absence of attributing any fine-grained structure to the system.

A second reply is that mathematical physics actually provides us with a formal apparatus for capturing causal dependencies more fine-grained than those among entire time slices even within the context of field theories. This reply will be the focus of the Section 5 and will be further developed in Chapters 5 and 7.

Formally, interventions into a causal model exhibit an asymmetry: interventions lead to changes in the values of variables causally downstream from the variable intervened into, but not to changes causally upstream from the intervention. Just such an asymmetry also characterizes our experimental interactions with physical systems. At least in the kinds of contexts we are familiar with, the causal asymmetry lines up with a temporal asymmetry in that effects do not precede their causes. Correspondingly, experimental interventions into an otherwise closed physical system affect the future evolution of that system but not its past, and this is so even when the dynamical laws governing the system are time-reversal invariant. This suggests the following argument for a role of causal representations in physics:

4.1   There is a temporal asymmetry characterizing experimental interventions into otherwise closed systems.

4.2   If there is such an asymmetry, it is best represented as a causal asymmetry.

4.3   Therefore, asymmetric causal representations play a legitimate role in physics.

Why should we accept 4.1? It is a striking fact about experimental interactions that we can only intervene into a system "from the past," as it were. Consider a system S that is governed by both past and future deterministic laws. That is, let us assume that the final state $S_f(t_f)$ of the system is uniquely determined by the initial state $S_i(t_i)$, where $t_i < t_f$, together with the dynamical laws and the boundary conditions, and that the initial state $S_i(t_i)$ is similarly determined by the final state $S_f(t_f)$. Thus, if S is closed between t_i and t_f, then the initial and final states mutually and symmetrically determine each other. Nevertheless, there is an asymmetry of state preparation in the following sense. We can prepare the system in its initial state $S_i(t_i)$ without making use of any knowledge we might have of the system's dynamical evolution between t_i and t_f, and we can also independently calculate the system's future evolution for times $t > t_i$ from the initial state, the dynamical laws, and the boundary conditions. But we cannot similarly first prepare the system's final state at t_f without using our knowledge of the dynamics and then take the final state together with the laws to calculate the system's past evolution for $t < t_f$.

Of course, we cannot first prepare the system in S_f and then let it *evolve* into S_i. This simply follows from the fact that S_f occurs after S_i. But that is not the asymmetry to which I want to draw attention. Rather, the asymmetry consists of the fact that we cannot *first* prepare the system in S_f *without making use* of the laws governing the dynamical evolution between S_i to S_f and *then calculate* what the system's past evolution from S_i to S_f must have been, given the dynamical laws and the boundary conditions. But we can prepare a system in a state S_i without appealing to the laws of evolution from S_i to S_f, and then use knowledge of the laws and boundary condition to calculate the final state S_f. That is, the asymmetry concerns a combination of state preparation, on the one hand, and subsequent prediction or retrodiction, on the other.

There are two ways in which we can prepare the system in a specific final state S_f at t_f. First, we can make use of our knowledge of the dynamical laws to determine the initial state S_i, in which the system has to start out at t_i in order to evolve into S_f, and prepare the system in the appropriate initial state S_i. In this case our ability to prepare the system in its final state S_f relies crucially on our knowledge of the laws and boundary conditions governing the system's evolution between t_i and t_f. This procedure can be used (and indeed has to be used) if the system is closed between t_i and t_f: if we cannot intervene into the system between t_i and t_f, we can only "prepare" it in the state S_f by preparing it in the earlier state S_i. And in order to know which state S_i will evolve into S_f, we need to make explicit use of the laws and boundary conditions characterizing the system. By contrast, even when the system is closed between t_i and t_f, we can still prepare the system in an initial state S_i without any knowledge of the dynamical evolution of the system between t_i and t_f. Although, given deterministic dynamical laws (plus boundary conditions), the initial and final states determine each other, we do not need to make use of that fact in setting up the system in some specified *initial* state.

We can imagine, for example, that one experimenter is responsible for preparing a system S in an initial state S_i or a final state S_f and that a different experimenter is responsible for setting up the boundary conditions characterizing the system for the time period from t_i to t_f. If the first experimenter wants to prepare the system in a certain initial state, he can do so without knowing what boundary conditions the second experimenter chooses to set up. But if the first experimenter wants to make sure that the system ends up in a specific final state, he needs to know what the boundary conditions will be in order to make sure he prepares the system in the appropriate initial state.

A second way of preparing the system in a final state S_f is to prepare the system in that state directly by intervening into the system between t_i and t_f. In this case we do not need to make use of our knowledge of the evolution between initial and final times. But since the system did not remain closed between t_i and t_f, we can no longer use the dynamical laws and boundary conditions governing the closed system to retrodict the initial state at t_i. Thus, there is a way for an experimenter to prepare the system in S_f without knowledge of the system's evolution between t_i and t_f but at the cost of having to violate the boundary conditions through his intervention and thus losing any ability to retrodict the evolution of the system.

Thus, even systems that are governed by both past and future deterministic dynamical equations exhibit an asymmetry of state preparation. In the case of a system that is closed between t_i and t_f and for which initial and final states mutually determine each other (given the dynamical equations and boundary conditions), we can only prepare the system in a final state S_f by making explicit use of our knowledge of the dynamics and preparing the system in the corresponding initial state S_i. If, by contrast, we allow interventions into the system between t_i and t_f, we can, directly intervene into the state of the system just prior to t_f, but then lose our ability to use the state at t_f to retrodict the state at t_i. We can, however, directly intervene into the state of the system just prior to t_i and then predict the system's evolution until some later time t_f. We can, that is, only intervene into a system from its past.

The asymmetry of state preparation is a paradigm case of the causal asymmetry, as understood by interventionist accounts of causation. In particular, if S_f is an effect of S_i, then there are two ways by which one can intervene into the system to set S_f to a particular value: first, we can intervene into S_i, which in turn will affect the value of S_f; or, second, we can intervene directly into S_f through an intervention that might be representable in terms of a causal model satisfying Woodward's switch condition and might even be genuinely "arrow-breaking," thereby making it impossible to retrodict the value of S_i on the basis of the value of S_f. Thus, according to an interventionist account, experimental systems will exhibit an asymmetry of state preparation, if earlier states of the system are causes of later states. The asymmetry provides us with an empirical justification for modeling physical systems causally, even if the model is extremely coarse-grained and represents the state of a system at t in terms of a single variable S_t.

The asymmetry of state preparation need not, and very often will not, be reflected in the dynamical equations used to represent a system, but there are cases where the intervention asymmetry plays a direct role in the formalism used to represent a system. Again the LHC can provide us with an example of this. We have already discussed the beam control system that measures the state of the proton beam and adjusts the beam in response to these measurements in order to keep the beam parameters stable. One potential problem with feedback systems of this kind is that there is a time lag between the measurements of the beam properties and the response of the control system. If this time lag is ignored, the feedback system will respond to what at the time of intervention is already outdated information about the state of the beam, and the control will not be optimal. This problem can be addressed by adding a so-called predictor function (an example of which is the so-called Smith predictor), which uses the measured state of the beam at some time t to predict its state at a slightly later time $t + \Delta t$ and calibrates the response to the predicted output. Important for our purposes here is that the model of the feedback system is explicitly time-asymmetric: the measurement precedes the intervention into the system by the control mechanism. Measurement and intervention are linked by a time-asymmetric delay term, which is taken to be explicitly causal. As Steinhagen, for example, explains: "Due to the causality, the delay term cannot be inverted" (Steinhagen 2007, 169).[4] In this example, then, the asymmetry of measurement and intervention finds an explicit representation in the asymmetry of the delay term.

The temporal asymmetry of causation plays an important role in the model of the control system, independently of any fine-grained structure attributed to the controlled system – the system intervened into. As I will discuss in Chapter 6, the asymmetry characterizing control systems is related to a causal asymmetry at the heart of so-called linear response theory in physics, which models the response of a system to an external force or field – that is, to an intervention into the system. It is often taken to be one of the major advantages of linear response theory that it allows us to make empirically useful predictions about the behavior of a system based on very general assumptions – such as a time-asymmetric principle that is usually identified as "principle of causality" – without having to provide a detailed model of the internal structure of the system.

5. The causal Green's function

In the last section I argued that causal representations can play a useful role even if we do not attribute a causal structure to a system any more fine-grained than that positing a causal connection among the total states of a system at different times. In this section I want to argue that mathematical physics does in fact provide us with a machinery that allows us to answer questions concerning causal dependencies more fine-grained than those between entire time slices. Woodward suggests that the existence of distinct causal routes is a consequence of adopting a coarse-grained perspective: "the whole notion that one variable might affect another via multiple distinct routes is itself a consequence of our adoption of a coarse-grained perspective and the distinctness of different routes itself disappears at a fine grained level" (Woodward 2007, 96, fn. 26). At a fine-grained level, an interventionist framework that relies on the ability to isolate and track how the effects of an intervention percolate through a system seems to be inapplicable. Yet, as I want to argue now, the notion of distinct causal influences can be captured even within the context of a field theory – that is, even for a theory that allows for continuous, non-discrete objects – even

---

[4] In a related context, Steinhagen says earlier: "It is obvious that causality forbids the inversion of the exponential delay term" (2007, 51)

though the notion of a discrete causal path becomes inapplicable within that framework.

The machinery I have in mind is that of the Green's function or so-called fundamental solution of a linear differential equation. Any linear differential operator L associated with an inhomogeneous differential equation $Ly = f(x)$ and with constant coefficients possesses a *fundamental solution* or *Green's function G*, which is a solution to the inhomogeneous differential equation $LG = \delta(x)$. In the case where we are interested in dynamical equations governing the values physical quantities take at spacetime points, the equation will take the following form: $LG(x, t, x', t') = \delta(x- - x')\delta$ $(t - t')$. The delta function $\delta(x)$ is a generalized function that (characterized somewhat informally) is zero everywhere except at $x = 0$, where it has the value 1. Physicists often use causal language in discussions of Green's functions. In fact, as Sheldon Smith points out, "discussions of Green's functions are a primary locus for causal claims within physics texts" (Smith 2007, 667). The Green's function is quite naturally interpreted in interventionist terms. The function "propagates a point inhomogeneity," as it were, and thereby tells us what the contribution of introducing a disturbance or perturbation into a system at (x', t') is to the state of the system at some other point (x, t). The Green's function formalism allows us to determine how different contribution to the effect at (x, t) add up. That is, the formalism allows us to represent the state at (x, t) as sum of a number of different disturbances as its causes. Moreover, the formalism even allows us to determine unperturbed causal dependencies between points in the system.

Now, it is sometimes argued that the causal significance of the Green's function formalism should not be overstated, since, at least in the case of hyperbolic equations, such as the wave equation, the formalism allows us to represent one and the same system in terms of both "causal" Green's functions and "anti-causal" Green's functions. As an inhomogeneous equation, a solution to the equation $LG = \delta(x)$ is unique only up to the addition of a solution to the corresponding homogeneous – that is, source-free – equation $LG = 0$. One solution is the so-called retarded or causal Green's function, which satisfies $F_{ret}(x, t, x', t') = 0$ for $t < t'$ and according to which a disturbance introduced at (x', t') does not affect the state of the system at times earlier than t'. But another solution is the advanced Green's function, which represents the state of the field as depending on the later state of the source. The advanced Green's function seems to represent disturbances that travel anti-causally backward in time. Other solutions can be constructed from linear combinations of the retarded and advanced

Green's functions. Applied to the case of a field theory, any total field F_{total} can be represented either as a sum of retarded Green's functions and a free incoming field F_{in}, which is a solution to the homogeneous field equations, or as a sum of advanced Green's functions and a free outgoing field F_{out}. That is, $F_{total} = F_{in} + G_{ret} = F_{out} + G_{adv}$. For this equality to hold, we just have to "carefully" choose the outgoing field to ensure that $F_{out} = F_{in} + G_{ret} - G_{adv}$. Both the retarded and the advanced representations are representations of one and the same solution to the dynamical equations. The former is a solution to an initial-value problem, whereas the latter is the solution to the corresponding final-value problem. Which representation we choose is up to us. Often the retarded or causal representation might be more convenient, but this does not give it any special significance, or so one might argue. That there are two kinds of representations of one and the same system that intuitively suggest a causal and anti-causal interpretation, respectively, might suggest that we should be careful in attaching too much significance to a causal reading of the formalism.

We will see in the next chapter, however, that there are reasons for privileging the retarded Green's function as that fundamental solution that gets the causal structure right. The argument for the equivalence between the two representations appeals to their equivalence as solutions to a full-fledged initial- (or final-) value problem. But very often, in modeling actual systems, do we not possess enough data to set up a full initial-value problem. This, as we will see, results in an underdetermination problem that can only be solved with the help of causal assumptions, which single out the retarded Green's function as causally privileged representation of the contribution of a point disturbance to the total field.

For now I want to posit without argument that the retarded Green's function gets the causal structure right. The machinery can then be readily incorporated into Pearl's do-calculus. The effect of an intervention at (x_i, t_i) that changes the state of a source, that is, $do(\rho(x_i, t_i))$, is, according to my proposal, determined with the help of the causal Green's function. Consider a system represented by a field theory with sources. A causal representation of this system provides a representation of the state of the system at some point (x, t) in terms of the state on some initial and boundary surface together with the retarded Green's function associated with any sources. That is, the causal representation tells us how the contributions of points on the initial and boundary surfaces and of source points add up to give the overall effect at (x, t). We can then ask where and how the state of the system would change if we introduced a disturbance at (x', t') that changed the field values at that point. The answer to this question is given by the causal

Green's function. Thus, Pearl-style counterfactuals of the form "$Y(x, t)$ would have value **y**, if $X(x', t')$ had value **x**" receive a unique and definite truth value if interventions are modeled with the help of the causal Green's function.

Smith has argued that the Green's function framework does not sit well with the causal covering law thesis, according to which if events are related as cause and effect, then they have descriptions that instantiate a causal law. The problem for the covering law thesis is that, as Smith argues, the Green's function is not in general an instance of the law with which it is associated, because of the delta-function singularity of the function. For example, "the Green's function for the wave equation [describing a vibrating string] *violates* the differential wave equation which is supposed to govern the *interior* region of the string" (Smith 2007, 674). But although this may spell doom for the covering law thesis, it is not a problem for the proposal I am considering here. Causal dependencies, I am suggesting, are given by the Green's function independently of whether the Green's function is an instance of the law with which it is associated.

We can restate in terms of the machinery of Green's functions the worry discussed in Section 3 that interventions into "large worlds" are ill-defined. If we ask what the effect of an outside disturbance on a system is, then this question, according to the objection, does not have a well-defined answer. If we model the disturbance in terms of the retarded Green's function (and keep the initial conditions fixed), we get one answer. But if we model the disturbance in terms of the advanced Green's function (and keep the final conditions fixed), we get a different answer. Worse still, different linear combinations of retarded and advanced Green's functions give us yet more answers. My reply is that the causal or retarded Green's function is privileged and is the correct one to use in modeling the intervention. So far this is nothing but a bald assertion. But in the next chapter I will provide an argument in its support.

In this section and the last I have offered two different responses to Woodward's and Hitchcock's worry that an interventionist conception of causation is inapplicable to models that do not allow us to distinguish discrete paths of causal influence. Both of these responses need to be developed further. I have argued that introducing a causal asymmetry for systems to which we do not attribute a fine-grained causal structure is justified in light of an asymmetry of interventions. Yet can the notions of causation and intervention actually play a useful role in contexts in which we do not attribute any causal structure to the system intervened more fine-grained than the state of a system at a time? As I will argue in Chapter 6, the answer to this question is "yes." The causal asymmetry characterizing control systems, which I mentioned earlier, is related to a causal asymmetry at the heart of so-called linear response theory in physics, which models the response of a system to an external force or field – that is, to an intervention into the system. In fact, it is often taken to be one of the major advantages of linear response theory that it allows us to make empirically useful predictions about the behavior of a system based on very general assumptions – such as a time-asymmetric principle that is usually identified as the "principle of causality" – without having to provide a detailed model of the internal structure of the system.

I have then argued that many physical laws allow us to introduce precise relations of asymmetric causal influence through the mechanism of the causal Green's function. Yet given the time-reversal invariance of (most of) the dynamical equations, choosing the *causal* or *retarded* Green's function over the *anti-causal* or *advanced* Green's function may seem arbitrary and without justification. In Chapters 5 and 7 I will show how we can introduce causal structures in the paradigm case of a classical particle-field theory, classical electrodynamics, and argue that these causal structures play both an important inferential and an important explanatory role in the theory. Thus, both the case of linear response theory and the use of causal Green's functions in field theories and elsewhere are cases that, as I will argue, meet Woodward's condition of legitimacy for causal notions according to which "causal notions are legitimate in any context in which we can explain why they are useful, what work they are doing, and how their application is controlled by evidence."

6. Conclusion

In this chapter I have examined various arguments for the conclusion that an interventionist notion of causation is inapplicable or at least not readily applicable to physics. I have argued that all these arguments are unsuccessful. In the process I have distinguished several different notions of intervention: that of Pearl's do-calculus, which is arrow-breaking but does not require positing an external intervention variable, and various variants of Woodwardian interventions that hook one or more intervention variables to the variable intervened into. I distinguished hard or genuinely arrow-breaking interventions from soft interventions and from feedback control interventions. Causal relations can be modeled in terms of any of these notions of interventions, and all of them have their legitimacy in physics, where context will determine which notion might be the most appropriate.

The tight conceptual connections between the notions of cause and intervention are often cited as a core reason for the claim that causal notions have a "human face" and therefore sit ill with our established theories of physics. As in the previous chapter, however, we have seen that this putatively human-faced feature of causal representations can play a legitimate role in how physics represents the world as well.

CHAPTER 5

The temporal asymmetry of causation

1. Introduction

In the last two chapters I examined a number of arguments for the claim that causal notions cannot play a legitimate role in established theories of physics. I argued that none of these arguments are successful. The general strategy these arguments employ consists of pointing to one or several putative contrasts between causal relations and the kind of structures presented to us by the theories of physics and to argue that the existence of these contrasts undermines the applicability of causal reasoning to physics. Perhaps the most telling such contrast is widely considered to be that between the asymmetry of the causal relation and the time-symmetric character of the dynamical laws of our established physical theories. In this chapter I want to examine whether this particular contrast can be forged into a successful argument to show that there is no place for causal representations in physics and will argue that arguments appealing to the asymmetry of causation are no more successful than the anti-causal arguments examined in the preceding chapters.

In the next section I discuss a preliminary argument that, although it does not itself invoke the asymmetry of the causal relation, is sometimes entangled with appeals to a causal asymmetry. This argument maintains that the fact that physical theories centrally involve abstract mathematical structures on its own already implies that these theories do not, or perhaps even cannot, also involve causal relations. In Section 3 I examine an argument due to Norton that aims to show that asymmetric causal relations are incompatible with the time-symmetric laws of physical theories. Norton has claimed that the combination of a time-asymmetric principle of causality and time-symmetric laws gives rise to a reductio ad absurdum. Yet the reductio argument fails. In Section 4 I will discuss the weaker view that although causal notions are not strictly incompatible with time-symmetric laws, there is nevertheless no scientifically legitimate reason for incorporating causal notions to physical theories. Huw Price and Brad Weslake, for example, argue that causal relations in physics would be practically irrelevant and epistemologically inaccessible. Against this argument I show that there are empirical reasons for positing causal structures. Limitations to the empirical evidence we have available for drawing inferences from one time to another result in an underdetermination problem that can only be solved with the help of causal assumptions.

2. Formulas and state-space models

I want to begin my discussion with an argument suggested by Ernst Mach and Bertrand Russell for the thesis that imprecise commonsense causal regularities are, in physics, replaced by precise laws that have the form of functional dependencies (see Chapter 1). Putatively causal claims, according to the argument, need to be underwritten by universal causal regularities of the form "All events of type A are followed by events of type B." But in trying to find such regularities, we are faced with the following dilemma. The events in question may be specified only vaguely and imprecisely. In this case the resulting regularities might be multiply instantiated, but they are formulated too imprecisely to be properly scientific. Or the events in question may be specified precisely, but then the resulting regularities are instantiated at most once. According to Mach and Russell, physics avoids this dilemma by providing us with precise functional dependencies between variables representing properties of event types – functional dependencies that can be both precise and multiply instantiated. Thus, the argument concludes, in physics the notion of cause has been replaced by that of functional dependency. But the conclusion does not follow. It does not follow from the fact that physical theories present us with functional dependencies rather than with simple regularities of the form "all A are B" that these functional dependencies themselves cannot be causal dependencies. An additional argument is needed. How, then, might we try to establish Mach's and Russell's conclusion?

During his discussion of Newton's law of gravity, Russell says that "in the motion of mutually gravitating bodies, there is nothing that can be called a cause and nothing that can be called an effect; *there is merely a formula*" (141, my emphasis). Decades later Bas van Fraassen echoes this claim, when he answers Nancy Cartwright's question (Cartwright 1993) "Why not allow causings in the models?" as follows:

To me the question is moot. The reason is that, as far as I can seen, the models which scientists offer us contain no structure which we can describe as putatively representing causings, or as distinguishing causings and similar events which are not causings... Some models of group theory contain parts representing shovings of kid brothers by big sisters, but group theory does not provide the wherewithal to distinguish those from shovings of big sisters by kid brothers. The distinction is made outside the theory. (van Fraassen 1993, 437–8)

While Russell's remark suggests that a theory ought to be strictly identified with a set of formulas, van Fraassen argues that a theory consists of a set of state-space models. But even though they disagree on whether theories ought to be understood syntactically or semantically, they agree that there is no place for causal notions in physics. According to van Fraassen's view of scientific representation more generally, a scientific theory presents us with a class of abstract mathematical structures that we use to represent the phenomena. These structures, van Fraassen suggests, cannot be used to represent causal relations – we cannot describe these structures as putatively representing causings – and do not allow us to draw a distinction between causal and non-causal relations. One way to read Russell's and van Fraassen's remarks is as claiming that it is the abstract mathematical nature of physical theories that renders physics inhospitable to causal notions.

Thus, we might try to reconstruct Russell's and van Fraassen's suggestions in terms of the following explicit argument:

2.1 The content of a physical theory is exhausted by a set of formulas or state-space models.

2.2 Causal relations are not part of the formulas or state-space models of a theory.

2.3 Therefore, causal relations are not part of the content of physical theories.

One might think that (2.2) is false (or at least not obviously true) and that causal relations can be part of a model. After all, in many disciplines scientists speak of "causal models" or "causal equations." But according to van Fraassen's view, the models presented by a scientific theory are, in the first instance, uninterpreted abstract mathematical structures – structures that acquire a representational role only when they are used or taken by us as representing certain phenomena. A causal model or causal equation, on this view, could only be causal insofar as the model or equation is used to represent what are taken to be causal relations.[1] The point might

---

[1] Just as models of $\mathbf{F} = m\mathbf{a}$ are not intrinsically models of Newtonian systems but only if we use $\mathbf{F}$, m, and $\mathbf{a}$ to represent force, mass, and acceleration, respectively.

be obscured by van Fraassen's use of the term "model," which has several distinct meanings both in science and in the philosophy of science, but (2.2) is simply a consequence of the claim that the core of a physical theory consists of abstract mathematical structures, which *on their own* are uninterpreted and do not represent anything. Put in terms of Russell's syntactic framework, at the core of the theory there is "merely a formula."

If understood as referring to uninterpreted formulas or mathematical structures, (2.2) appears to be true, but under this disambiguation (2.1) is obviously false. No theory of physics can be strictly identified with a set of formulas or uninterpreted state-space models, because in order to make any claims about the world, the theory must contain an interpretation that tells us which bits of the formalism are hooked up with which bits of the world. Minimally, a theory's interpretation has to specify the theory's ontology – it has to specify which parts of the world the various components of the mathematical structures are intended to represent. But once we see that the austere view of theories as consisting solely of a mathematical formalism or set of abstract mathematical structures is untenable and that an interpretive framework needs to be part of a theory, the question arises: why could this framework not be rich enough to include causal assumptions as well? For example, an interpretive framework for Newton's laws might not merely specify that m represents mass, F force, and a acceleration but might include the causal assumption that forces are causes of accelerations. That is, we cannot conclude from the fact that an *uninterpreted* formula $F = ma$ does not on its own mark F as cause and a as effect – that the causal "distinction is made outside the theory."

Thus, our first attempt at distilling a successful argument out of Russell's and van Fraassen's remarks failed. A second suggestion is that there might be constraints on what can be part of the formalism's interpretation that exclude causal notions. Thus, Earman has proposed that a theory's content is exhausted by a formalism together with what he calls a "minimalist interpretation" (Earman 2011, 494). Thus, we should replace (2.1) with the following claim:

> 2.1′ The content of a physical theory is exhausted by a set of state-space models or a set of formulas *together with a minimalist interpretation.*

But instead of answering the question as to what is allowed to be part of the theory's interpretive framework, this proposal merely postpones the question. What can properly be part of the minimalist interpretation, and on what grounds can causal interpretations of certain mathematical relations be excluded? Earman's suggestion, echoing Russell's view, is that a minimalist interpretation is one that is free from "philosophy-speak" (505) and, thus, cannot involve the notion of cause. But this still does not provide an argument in support of the causal skeptic, for what is lacking is an account of what distinguishes "philosophy-speak" from legitimate "physics-speak." The criterion cannot be to exclude notions that are employed by philosophers but not by physicists, because the physics literature is replete with appeals to causality – for example, as "physically well-founded assumption" (Jackson 1975, 312), as "fundamental assumption" (Nussenzveig 1972, 4) or as "general physical property" (Nussenzveig 1972, 7), or even as the "most sacred tenet in all of physics" (Griffiths 2004, 424).

Van Fraassen's remarks quoted earlier suggest an alternative way of spelling out the idea of a minimalist interpretation: only those terms are a legitimate part of the interpretive framework that correspond to a part of the formalism: for each physical correlate of the mathematical models posited in the interpretation, we have to be able to identify the element or substructure in the models that represents that part of the world. The anti-causal claim then is that there are no substructures in the mathematical models presented to us in physics that can be described as representing causings. For example, one might ask where in the state-space models defined by Newton's laws we find anything that could be taken to represent causal relations. What is more, van Fraassen even appears to suggest that causal relations *cannot* be represented structurally in a mathematical model, for he says that "group theory does not provide the wherewithal to distinguish" asymmetric causal relations from their inverses or, in his example, shovings of kid brothers by big sisters from shovings of big sisters by kid brothers. Asymmetric causal distinctions, therefore, would have to be drawn outside a theory.

To take the second worry first, formal work on causal models such as Pearl's structural theory of causation shows that causal assumptions can be represented mathematically. In van Fraassen's toy example, we can define an asymmetric relation R over the domain of objects consisting of all sisters and brothers, which we interpret as the "a shoves b" relation, and there will be models in which some a that are sisters stand in relation R to some b that are brothers, but in which no brothers stand in relation R to any sisters. In these models it will be true that some sisters shove their brothers, but it will not be true that any brothers shove their sisters. Thus, group theory or mathematics more generally does appear to provide the wherewithal to distinguish shovings of sisters by brothers from shovings of brothers by sisters.

One might reply that all that the mathematical formalism allows us to do is to define an asymmetric relation, but the formalism itself does not allow us to distinguish between the relation "x is a *cause* of y" and the relation "x is an *effect* of y": the formalism alone does not distinguish between the "shoves" and "is shoved by" relation – and perhaps this is van Fraassen's point. But just as the formalism on its own cannot determine which objects its different variables represent and only represents certain objects in virtue of it being *used* to represent these objects, our use can equally determine that a given asymmetric relation represents the "cause" rather than the "effect" relation. The distinction is not made outside of the theory but in the theory's context of use.

What remains is van Fraassen's first worry: although there might be structures that we "*can* putatively describe as representing causings," one might insist that these structures are not part of the models scientists do, *as a matter of fact*, use. To present a theory is simply to present a class of (suitably interpreted) state-space models, defined by a theory's basic equations. But it is not obvious to me that this last claim is correct. As the foregoing quotes suggest, physicists often do invoke causal assumptions. And such informal appeals to causal principles could be understood as implicitly defining causal structures into which models of the dynamical equations are taken to be embedded. Since the causal structures in question often are very simple, little or nothing might be gained from adding a formal representation of these structures to the theory's equations. Nevertheless, if we want to offer a formal philosophical reconstruction of a theory, say, along the lines of van Fraassen's semantic view, we would have to include a representation of any causal assumptions, even if physicists themselves never represent these assumptions formally in terms of a partial ordering relation. Of course, to establish whether physicists' explicit appeal to causal assumptions in any particular theory ought indeed to be understood as a commitment to causal structures requires a detailed case-by-case investigation. My present point is merely that we cannot conclude simply from the fact that the models *of a set of equations* do not contain structures representing asymmetric causal relations that *scientific theories* contain no asymmetric causal assumptions and that any causal "distinction is made outside the theory."

If we want to establish that causal relations cannot be part of how physical theories represent the world, we need a more substantive argument than merely an appeal to the formal character of functional dependencies or state-space models. What exactly is it about the mathematical machinery of physical theories that appears to render causal notion incompatible, or at least makes them sit ill, with physical theories? As I said earlier, the most prominent suggestion in the literature is that it is the time-reversal invariance of the dynamical equations, which is incompatible with asymmetric causal relations playing a legitimate role in physics.

3. Time-symmetry

The contrast between the time-reversal invariance of the dynamical laws and the time-asymmetry of the causal relation can be fashioned into an explicit anti-causal argument as follows:

3.1 Causal relations are temporally asymmetric.

3.2 The physical laws of our well-established theories have the same character in both the forward and backward temporal directions.

3.3 Therefore, there is no place for time-asymmetric causal relations in a theory with time-symmetric laws.

3.4 Therefore, there is no place for the causal relations in our well-established theories of physics.

Both premises (3.1) and (3.2) would deserve further comment: (3.1) appears to deny the possibility of instantaneous causation, while (3.2) might strike one as obviously false: there are many well-established but non-fundamental theories that are not time-symmetric, and there are even arguably fundamental theories that are not time-reversal invariant.[2] But let us restrict our attention to well-established theories that are not explicitly phenomenological, as thermodynamics is, and follow the perhaps unjustified practice of ignoring failures of time-reversal invariance in particle physics. Thus, here I want to focus on the inference from premises (3.1) and (3.2) to (3.3).

One might read this inference as relying on the same assumptions about the content of a physical theory as the argument in the preceding section – the assumption that the content of a theory is exhausted by a set of state-space models with a minimal interpretation that associates the "mathematical squiggles" of an equation with physical quantities. This appears to be how Alyssa Ney construes this argument in her convincing criticism of it (see Ney 2009, 748).[3] Alternatively, one might take the appeal to the *temporal character* of our laws as crucial to the argument. The thesis then

---

[2] At this point in the discussion many philosophers note, but then dismiss, the fact time-reversal invariance seems to be violated for elementary quantum particles. See Maudlin (2007) for a criticism of this practice.

[3] Ney (2009) also argues for a place for causal notions in physics, and I find many of her arguments convincing. Where we part company is concerning her conclusion that a notion of cause appropriate for physics needs to be symmetric. My argument here is that a time-asymmetric notion of cause plays an important role in physics.

is, that although in principle causal structures could be part of a theory, the fact that a theory's laws are in some sense time-symmetric prohibits this.

The claim that the laws have the same character in both directions is vague and allows for two distinct readings, as Alexander Reutlinger and Matthew Farr also point out (Farr and Reutlinger 2013). The first reading, which is suggested by Russell's remark that the laws make no difference between past and future in that "the future 'determines' the past in exactly the same sense in which the past 'determines' the future," states that the laws in question are both past and future deterministic. The second reading states that that the laws of physics are time-reversal invariant. These two readings result in distinct arguments.

Here is the argument from determinism:

3.1.D   The fundamental equations of physics are both past- and future-deterministic.

3.2.D   There is no place for an asymmetric notion of cause in the context of a theory with fundamental equations that are both past- and future-deterministic.

3.3.D   Therefore, there is no place for an asymmetric notion of cause in mature physical theories.

Of course, we no longer believe that the fundamental laws of nature are deterministic, but there are independent reasons for rejecting the conclusion. Premise 3.2.D implies that in situations where causes determine their effects, the set of effects of an event cannot in turn determine its causes, and this premise does not appear to be defensible. Why should it be incompatible with the very notion of a causal relation, that a set of complete effects of an even C cannot be nomologically sufficient for C's?

Consider Pearl's structural account of causation, according to which a causal model contains structural equations $x_i = f_i(pa_i, u_i)$, which determine the values of the effect variables in terms of the values of their causal parents. According to the argument from determinism, there is a constraint on the form any such structural equations can take: the complete set of equations $x_i = f_i(pa_i, u_i)$ must not be solvable for the causes pa_i as functions of their effects x_i. Recall that the equality sign in a structural equation is not the familiar symmetric equality sign. In a structural equation, the effect is always to the left of the equality sign and is expressed as function of its causes that stand on the right-hand side of the equation. That is, qua structural equations, the equations cannot be inverted. But what the argument from determinism implies is that the functional dependencies in any causal model have to be such that, if we replaced the set of *structural* equations with analogous regular equations with the standard symmetric equality sign, we could not solve the resulting expressions for a set of causal parents pa_i as functions of their effects x_i. For example, the argument from determinism implies that it is *conceptually impossible* to have the following set of structural equations: $x = u + v$; $y = u + w$; $z = v$. Since the regular-equation analogues of these equations can be rewritten as $v = z$; $u = x - z$; $w = y - x + z$, the causes u, v, and w are determined by their effects x, y, and z in violation of the argument from determinism. It is difficult to see what a non-question-begging justification for such a restriction on the structural equations might be.

Or consider the metaphysical claim that causes produce or bring about their effects. Whatever the merits of such a metaphysical view of causation ultimately might be, it is difficult to see why the claim that causes *produce* their effects would preclude the possibility that effects might be *nomologically sufficient* for their causes. The notions of nomological determination and causal production must be carefully distinguished, and it is unclear why a world in which effects jointly nomologically determine the causes that produce them should be conceptually impossible.[4]

The second reading of the incompatibility argument appeals to the time-reversal invariance of a theory's fundamental equations and can be expressed as follows:

3.1.R The fundamental equations of all mature physical theories are time-reversal invariant.

3.2.R An asymmetric notion of cause is incompatible with time-reversal invariant laws.

3.3.R Therefore, there is no place for an asymmetric notion of cause in mature physical theories.

Premise (3.2.R) is often asserted without offering much of an argument in support of it. For example, Scheibe simply concludes, after pointing to the contrast between time-symmetric laws and time-asymmetric causal relations, that "this suffices to seal the fate of event-causality" (Scheibe 2006, 213).[5] One of the few exceptions is an argument by Norton that aims to show that one can derive a contradiction from the conjunction of time-symmetric dynamical laws with a time-asymmetric asymmetric causal assumption. Norton's argument occurs in his reply to my (2009) discussion

---

[4] Thus, even though I find much of their discussion illuminating, I disagree with Reutlinger and Farr that the determinism-version of the temporal asymmetry argument is successful (see Farr and Reutlinger, 2013).

[5] "Schon [mit diesem Kontrast] scheint mir das Schicksal der Ereigniskausalität als fundamentaler Gesetzlichkeit besiegelt zu sein." (The translation into English is my own.)

of the role of causal assumptions in the derivation of dispersion relations. I want to quote Norton's argument in full:

> Now imagine a universe completely empty excepting two processes that we will call 'A' and 'B'. Process A has an incident wave, a dielectric, and a scattered wave. Process B is the time reverse of A. The two processes are completely isomorphic in all properties. Any property of one will have its isomorphic correlate in the other. Any fact about one will have a correlate fact obtaining for the other. One might be tempted to imagine that one of the two processes is 'really' the ordinary one, progressing normally in time; while the other is a theoretician's fantasy, a possibility in principle, but in practice unrealizable. The essential point of the example is that no property of the A and B systems distinguish [sic] which is which. Every property of one has a perfect correlate in the other. Let us assume that Frisch's principle of causality applies to one of these processes, the A process, for example. That will be expressed as a condition that the present state of the process depends only on its past states. Exactly what 'depends' may amount to is to be decided by the principle. All that matters for our purposes is that an exactly isomorphic condition of dependence will be obtained in the B process, except that it will be time reversed. Indeed, using the time order natural to process A, we would have to say that the principle of causality requires the present states of process B to depend upon its future states. In short, if the principle applies to process A, it fails for process B; and conversely. This is a *reductio ad absurdum* of the applicability of Frisch's principle of causality to scattering in classical electrodynamics. (Norton 2009, 481–2)

I will discuss the specific example to which Norton refers – derivations of the classical dispersion relations for an electromagnetic wave incident on a dielectric medium – in much more detail in the next chapter. The topic of that chapter will be the role of causal assumptions in linear response theory, of which derivations of dispersion relations are a special case. In order to assess the merits of Norton's argument here, the details of the physical case do not matter.

The argument appears to be this. Let us begin by postulating time-symmetric dynamical laws that allow a certain process A to occur, which is time-*asymmetric*. Since the laws are time-*symmetric*, they also allow the time-reverse of A, the process B, to occur. If we then posit a causal principle according to which future states causally depend on past states (but not past states on future states) and which we assume A to satisfy, we can derive a contradiction: on the one hand, since A satisfies the causal principle but the dynamical laws are time-symmetric, B, in virtue of being the time-reverse of A, will satisfy an *inverse* causal principle according to which a *past* state of the process B causally depends on its *future* states (but not vice versa).

But on the other hand, since the causal principle is assumed to be general, B will also satisfy the original principle, and *future* states of the process should depend on the *past* state (but not vice versa). This concludes the reductio ad absurdum.

In explicit premise-conclusion form, the argument can be expressed as follows:

3.5 There is a time-asymmetric dynamical process A governed by time-symmetric dynamical laws.

3.6 B, the temporal inverse of A, is dynamically possible. (5)

3.7 A and its temporal inverse B have exactly the same physical properties. (5, 6)

3.8 For all processes, future states causally depend on past states (but not vice versa). (Causal Principle)

3.9 Future states of A causally depend on its past states. (8)

3.10 Past states of B causally depend on its future states (but not vice versa). (7, 9)

3.11 Future states of B causally depend on its past states (but not vice versa). (8)

That is, the conjunction of time-symmetric dynamical laws with a time-asymmetric causal principle results in a contradiction.

Premise (3.5) cannot be assailed, since even though we assume the laws to be time-symmetric, many – and in fact in some intuitive sense, most – models of the laws will be time-asymmetric. But a defender of a causal principle should resist the steps of the argument leading to (3.10), and in particular the inference from (3.5) to (3.7) and (3.10): It does not follow from the assumptions that B is the dynamical time-reversal of process A and that A satisfies a time-asymmetric causal principles that B will satisfy an inverse causal principle. Since Norton's inference might strike one as initially plausible, I want to be belabor this point a bit.

Let us assume that purely dynamical time-asymmetric models of A and B can be represented by non-directed graphs:

(A) and (B)

According to the causal principle, both models can be embedded into richer structures that include an asymmetric causal relation, which can be represented by adding a direction to the graphs:

(A_{causal}) (B_{causal})

Norton points out that there is nothing in the purely dynamical models (that is the mathematical structures satisfying the dynamical laws) that tells us which model is which: there is no intrinsic difference between the two dynamical models. But the symmetry between the two models is broken in the directed causal models. And since the principle of causality is a general principle, once we "add the arrowheads" to one graph, as it were, this fixes the direction of the arrows in the other graphs, as long as we assume that the relative temporal orientation of the different models is given. Whatever models are used to model two physical processes A and B, respectively, we know that their temporal orientations are opposite to each another.

According to Norton "the principle of causality requires" also that we represent the putatively causal process B by the inverse graph:

 $(B_{anti\text{-}causal})$

Norton's reason is that it follows from the fact that the two processes A and B are time-reverses of each other that there is no property that distinguishes them. Since there is no physical difference between the two processes A and B, whatever reasons we might have for adding arrows to the graph representing A, which are directed from the vertex of degree 2 – that is, the vertex with two edges at the bottom of the graph just shown – to the two vertices of degree 1, the very same reasons would imply that we have to draw arrows in the graph representing B from the vertex with degree 2 at the top of the graph to the two vertices with degree 1 at the bottom.

But this step in the argument begs the question against someone who believes that causal relations play a substantive role in physics and maintains that it is precisely the causal properties of the two processes that distinguish them from each other: In the causal process A, the event represented by the vertex of degree 2 causes the events represented by the two vertices of degree 1, whereas in the causal process B, the events represented by the two vertices of degree 1 cause the event represented by the vertex of degree 2. That is, (3.7) does not follow from (3.5) alone, but requires as additional assumption the claim that the physical properties of a physical system are exhausted by those captured in the dynamical equations governing the system. But this is precisely the assumption a defender of causal relations wishes to deny, who would want to insist that the arrows in the causal structure also represent features of the system – features that cannot be derived from the dynamical equations alone.[6] Whereas there is no difference between the two purely *dynamical* models represented by the two non-directed graphs, there is a difference between the two physical processes represented, a defender of a causal principle would insist, and that difference consists of the difference in *causal* structure represented in the two directed graphs A_{causal} and B_{causal}. Thus, the attempted reductio fails.

A defender of causality in physics insists that causal relations, which are not implied by the purely dynamical properties of a system also play an important role in the representation of physical systems. The causal skeptic denies this and maintains that any appeal to asymmetric causal structures in addition to a theory's purely dynamical models is unfounded and scientifically unjustified. Thus, in order to complete Norton's anti-causal argument, we would have to add as additional premise the claim that positing causal structures is unjustified. But with this additional premise, it becomes unclear what the overall structure of the argument is meant to be. The premise would itself have to be supported by an argument, but such an additional argument would, if it could be made successfully, render the reductio proposed by Norton superfluous. If one were able to show that there are no scientifically legitimate reasons for positing a physical difference between processes A and B, the causalist would be defeated and there would be no work left to be done for the reductio argument. Thus, without an additional argument for the implicit premise in the argument from (3.5) to (3.11), the reductio begs the question against the causalist, but if we had such an additional argument, that alone would suffice to make the case against the causalist.

4. A problem of underdetermination

In the previous section we examined an argument for the claim that time-asymmetric causal assumptions are incompatible with physical theories with time-symmetric dynamical laws. We saw that the argument is unsuccessful. The argument failed because it had to assume what a defender of a "causal principle" would want to deny – that there are no legitimate reasons for positing causal relations that go beyond what is implied by a physical

---

[6] Even though I am following Norton here in expressing the argument in terms of real physical properties of a process, the point I wish to make here is independent of the debate about scientific realism. A defender of a principle of causality can also be an instrumentalist and argue that the causal relations in our models no more represent real properties than other properties or relations in our models. My claim here is that there is no principled reason for treating causal features differently from other kinds of properties and relations of our models (and of the real-world systems we are modeling).

theory's dynamical equations. A perhaps more promising argumentative strategy is to focus on this last claim directly, and to try to argue that, while causal claims might not be strictly incompatible with the time-symmetric laws of physics, the use of causal notions cannot be justified within the context of such laws and can be no more than a scientifically unmotivated add-on. Thus, the claim is that it would be a mistake to accept a causal principle not because it is strictly incompatible with time-symmetric laws but because *there are no good reasons* for positing causal structures in addition to the non-causal properties represented in the dynamical laws. The basic equations of a theory that is future- as well as past-deterministic define both an initial and a final-value problem. If we begin with the system's initial state, then the dynamical equations determine the system's subsequent evolution; if we take the system's final state to be given, then the dynamical equations determine the system's earlier evolution. Different models are distinguished by different initial and boundary conditions. Once we are given the dynamical equations in conjunction with appropriate initial conditions, there appears to be no work left to be done by putatively causal principles: the laws plus initial conditions tell us everything there is to know about the system in question. What is more, if the laws are time-symmetric, there appears to be no reason for distinguishing one temporal direction as the direction of causal influence. Thus, one might be tempted to agree with Fritz Rohrlich, who once maintained that the "identification of *causality* with *prediction* rather than *retrodiction* in a time-symmetric system of equations is completely arbitrary" (Rohrlich 2007, 51, italics in original).[7] While there may be no good arguments that strictly *disallow* interpreting a theory causally, it might nevertheless be the case that there could be no scientifically legitimate reasons for supporting an asymmetric causal interpretation of a theory.

This view is forcefully defended by Huw Price and Brad Weslake, who conclude from the premise that "fundamental physics seems to be time-symmetric" (Price and Weslake 2009, 416) that if time-asymmetric causal relations were to be real, they would have to be something "over and above physics" (417). In light of the fact that dynamical models already give us a complete representation of the temporal evolution of a physical system, a defender of causal relations would have to resort to a "hyperrealist view of causation" – a view that is deeply problematic: "the main difficulty with hyperrealism is that in putting causation beyond physics, it threatens to

---

[7] Rohrlich appears to have changed his mind and in later work suggests that there can be good reasons for interpreting a theory with time-reversal invariant laws asymmetrically causally. (See, e.g., Rohrlich 2006.)

make it both *epistemologically inaccessible* and *practically irrelevant*" (417, italics in original). In a physics with time-symmetric laws that pose a well-defined initial-value problem, there can be no empirical justification for positing causal relations.

Again we can represent the argument in explicit premise-conclusion form:

4.1 Reasoning and inferences in physics can be exhaustively characterized in terms of a theory's dynamical models together with choices of particular initial and boundary conditions.

4.2 Time-symmetric equations cannot provide evidence for asymmetric causal assumptions.

4.3 Asymmetric causal notions could play a legitimate and substantive role in a physical conception of the world only if either they played a substantive role in explanations or inferences in addition to the purely dynamical models or their use was justified by the character of our theories' dynamical laws.

4.4 Therefore, asymmetric causal notions can play no legitimate role in a physicalist conception of the world. (1, 2, 3)

This argument is an Ockham's razor argument: physics has no need for asymmetric causal relations and therefore one should not posit such relations. Yet both premises (4.1) and (4.3) can be challenged.

First, contrary to what (4.1) asserts, many (and arguably most) inferences in physics do not proceed from fully specified initial conditions fed into the appropriate dynamical equations. Rather, in many cases our observational data severely underdetermine which purely dynamical model of a given theory adequately represents the phenomena. This underdetermination problem is (at least often) solved with the help of causal structures. Thus, causal assumptions also play an important *inferential role* in physics and are not practically irrelevant.

Second, in order to establish that causal relations would have to be extra-physical relations that can play no legitimate role, it is not enough to point to the temporal symmetry of the laws. One would in addition have to show that time-asymmetric causal assumptions do not play a role in our treatment of initial or boundary conditions. In fact, there is a stark asymmetry between prevailing initial and final conditions. Initial but not final conditions satisfy a randomness assumption, and this asymmetry can justify the use of causal assumptions, as I will argue later.[8] Thus, causal

---

[8] See Maudlin (2007, 120) for a similar argument.

assumptions play an *explanatory role* even in the context of full dynamical models and are not epistemologically inaccessible.

I will examine (4.1) in this section and will turn to (4.3) in the next section. The underdetermination problem and its solution are quite general, but I want to illustrate it in terms of a particularly stark example: inferences from astronomical observations. Imagine you are looking up at the night sky and are observing points of light. How can we scientifically justify our belief that these points are the light emitted by stars? What licenses our belief that a given light point was indeed emitted by a star as its source rather than being source-free radiation coming in from past infinity? It appears to be almost religious dogma among many philosophers of physics that the content of a physical theory is exhausted by the models of its dynamical equations and that the only way to use a theory to make empirical predictions is to solve an appropriate initial- or final-value problem. I already questioned this dogma in Chapter 3 and now will give further reasons for its inadequacy. If we wanted to use the machinery provided by the dynamical laws governing the emission and propagation of light – the Maxwell-Lorentz equations and the wave equation that can be derived from them – to determine whether a locally observed light point is associated with a star as its source, we would have to solve a final-value problem: we would have to determine the state of the world – that is, the values of the field and the state of motion of any sources – on a final-value surface, which we could then feed into the dynamical laws and evolve backward in time to the state of the world at the location P of the putative source. Because the dynamical equations at issue are relativistic equations, the final-value surface required to determine the state of the world at the putative source point P is a spacelike cross section of the entire forward lightcone centered on P (see Figure 5.1) – that is, a spatial sphere that has a diameter of many light years, depending on the star in question! But obviously we do not have access to the required data: the only data we have at our disposal for inferring the existence of the star are our highly localized observations of the electromagnetic fields here on Earth.

In fact, there is an additional problem, if we use classical electrodynamics to infer the existence and the state of the putative source of a light point. Because of the problem of self-interactions – the interaction of a source with its own field – radiation phenomena are usually modeled in terms of a modified initial-value problem in which the full trajectories of the sources are assumed as given and the time-evolution of the fields is calculated from these and the fields on an initial-value surface (see Frisch 2005a for a detailed discussion of this problem). Yet neither do we know the fields

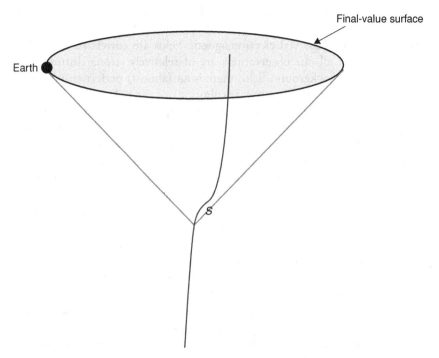

Figure 5.1 Cross section of a lightcone centered on the trajectory of the star

on anything close to a complete initial- or final-value surface, nor do we have independent access to the trajectories of the putative source – the star emitting the radiation. Thus, if the only theoretical tools at our disposal for making inferences about the putative sources of stellar radiation were the relevant dynamical laws applied to a pure or modified initial-value problem, it would be a complete mystery as to how we could ever be able to justify our belief in the existence of a star.[9]

Yet we do seem to be able to make justified inferences not only about the existence of stars but also about many of their properties based on the radiation fields we observe on Earth. What, then, is the structure of these inferences? The answer, it seems to me, is that our belief in the existence of a star as the source of the observed radiation is justified through a paradigmatically causal inference. There are strong correlations among the

[9] Indeed, in order to determine that the observed field is a radiation field (that is, a field associated with a source), we would have to know the field on the complete final-value surface. There is no local condition that is sufficient for the field to be a radiation field. (See Rohrlich 2007.)

light points observed at the same celestial latitudes and longitudes both at different times and from different locations on Earth at one time. In fact, the locally observed electromagnetic fields are correlated in several different ways: all our observations are of relatively strong disturbances in a very weak background field; there is an (almost) perfect coincidence in the luminosities and spectral distributions of the radiation observed at different spatiotemporal locations; and perhaps even more strikingly, the shapes of the field disturbances received at different spatial locations match so closely that they can be made to interfere with one another – a fact that is exploited in stellar interferometry. The degree of partial coherence in this last sense can be expressed in terms of so-called coherence functions associated with the fields (see, e.g., Born and Wolf, 1999, Chapter 10). Since the celestial latitudes and longitudes at which the correlated field disturbances at different times are observed are such that the fields can be associated with the trajectory of a single localized source in relative motion to us, we infer the existence of a star as common cause of our observations as providing the best explanation for the observed correlations.

Consider the following two dynamical models D_1 and D_2, which are solutions to the dynamical equations compatible with two observations of light points. D_1 involves only free fields coming in from past infinity, while D_2 posits a star as source of the observed radiation and contains free incoming fields that are approximately equal to zero. In fact there are many more models compatible with our observations, including models involving multiple sources and free fields that serve to mask the presence of these sources to ensure that the observed fields are the ones compatible with a single source in the presence of weak incoming fields, but we can focus on just D_1 and D_2 here. The point I want to stress is that both models are compatible with all the available evidence, and there is no reason provided by the dynamical models alone to prefer one model over the other.

Now contrast D_1 and D_2 with two corresponding causal models C_1 and C_2. As in Chapter 4, we can take a causal model C to consist of a set of variables $V = \{X, Y, \ldots\}$, a directed acyclic graph, and a set of structural equations F that specify the value of each variable in terms of the value of the variable's causal parents. C_1 contains non-zero source-free fields F_1 and F_2 at some time in the remote past as two independent causes of our observations O_1 and O_2 (see Figure 5.2). C_2 in addition contains a field source C and takes each observation O_i to be jointly caused by C together with any free incoming fields F_i as separate causes (see Figure 5.3). The two causal graphs on their own represent two different types of causal structure but do not specify a quantitative dependence of the variables O_1 and O_2 on their causes.

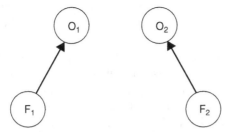

Figure 5.2 Causal model C_1

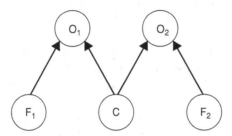

Figure 5.3 Causal model C_2

The quantitative dependence is given by structural equations, which can in this case be determined from the dynamical equations even without solving a full initial- or final-value problem: the selective dependence of the values of the observation variables O_i on each of its causes is given by the causal Green's function, which determines the effect on the total field of a pointlike disturbance located at (x', t').

As we have seen in Chapter 4, the Green's function G associated with a linear non-homogeneous differential equation $Ly = f(x)$ is defined as $LG(x, x') = \delta(x - x')$. If f depends on space and time variables, this becomes $LG(x, t, x', t') = \delta(x - x', t - t')$. As an inhomogeneous equation, its solution is unique only up to the addition of a solution to the corresponding homogeneous (i.e., source-free) equation. Demanding in addition that $G(x, t, x', t') = 0$ for $t < t'$ constrains the solution to the "causal" or "retarded" solution, which ensures that the value of a variable causally depends only on variables in its past.[10] Thus, the functional dependencies

---

[10] In the case of the wave equation, the causal solution is a wave coherently diverging from the source. Other solutions are a wave coherently converging on the source or any linear combination of these two solutions. These solutions are excluded by the causal condition that the source affects the total field only after the source is turned on.

of a variable on its causes in a causal model are given by the causal or retarded solutions to the fundamental equation associated with the dynamical law governing the system.

Even though both causal models C_1 and C_2, like their purely dynamical counterparts D_1 and D_2, are also in principle compatible with our observations O_1 and O_2, there is a causal principle that allows us to decide between the two models: the principle of the common cause (PCC), which states that if two variables X and Y are correlated with each other and are not related as cause and effect, then there is a third variable that is a common cause of X and Y. C_2, but not C_1, satisfies this principle. Thus, it follows from the principle of the common cause that C_2, rather than C_1 is the correct causal model into which to embed our observations, and it is this principle that licenses our inference to the existence of a star. What is the status of the common cause principle? There is a large literature on the principle, including discussions of a number of apparent counterexamples to the principle (see Arntzenius 2010 for a survey). Nevertheless, common cause reasoning clearly plays an important role in inferences toward the past, and the counterexamples can be avoided if we think of a principle of the common cause as a defeasible epistemological guide rather than as a metaphysical principle.

In somewhat more detail we can reconstruct the common cause inference in our case as follows. Consider, as a preliminary step, the field diverging from a radiating source. As the field spreads, its intensity decreases with $1/r^2$, where r is the distance from the source. Thus, the intensity of the field decreases with time. In the distant future, as the intensity of the field tends toward zero, the only remaining effect of the radiating source will consist of distant correlations among ever smaller and eventually infinitesimal field strengths. Now consider the temporal inverse of this process, which consists of correlations among infinitesimal fields in distant field regions in the remote past, which gain in intensity as they converge on and eventually collapse into the source. The further back in time we follow the field converging on the source, the weaker the field will be, becoming more and more dispersed toward the past and originating in what ultimately would seem to be extremely delicately coordinated infinitesimal correlations among distant field regions. Such correlations among initial fields strike us as extremely miraculous as even forceful opponents of the legitimacy of causal notions in physics, like Earman, emphasize:

> It would seem nearly miraculous if the time reverse of [the broadcast wave] were realized in the form of anti-broadcast waves coming in from spatial infinity and collapsing on the antenna. The absence of such near miracles

might be explained by an improbability in the coordinated behavior of incoming source free radiation from different directions of space. (Earman 2011, 506–7)

As Earman suggests, the intuition behind the idea that delicate correlations among the fields in distant regions would be nearly miraculous can be expressed probabilistically. If we allow infinitesimal fluctuations among the very weak fields in the remote past and assume that these fluctuations are random – that is, if we assume an equi-probability distribution over infinitesimal fluctuations in the field vectors – then the probability of the kind of coordinated behavior needed to result in a converging "anti-broadcast wave" is extremely small. Obviously, this probabilistic assumption is time-asymmetric: we find nothing miraculous in correlations "delicately set up" among infinitesimal fields in the distant *future* of a broadcast source, since these correlations have an explanation in the earlier action of the source as their common cause.

A source-free field mimicking the presence of a source, as in the models D_1 or C_1, is a field that collapses from different directions into the location of the putative source and then rediverges. Thus, a source-free model into which we embed our correlated observations of light points would have to contain exactly the same kind of infinitesimal correlations between distant field regions in the remote past as a field converging into a source. We can capture the assumption that coordinated behavior of incoming source-free radiation is improbable in the causal models by positing a probability distribution over the exogenous variables in the deterministic causal model C_2, which represent the free initial fields in the remote past, and assuming that these fields in different spatial regions are probabilistically independent of one another. Since the incoming fields in C_1 are highly correlated, the exogenous variables in C_1 are not independent. It can then be shown that C_2 (but not C_1) satisfies the causal Markov condition, which states that, for every variable X in V, X is probabilistically independent of the variables in the set $(V - \textbf{Descendents}(X))$ conditional on the parents of X. That C_2 satisfies the causal Markov condition follows from the causal Markov theorem, which says that any deterministic acyclic causal model with independent exogenous variables satisfies the causal Markov condition (for proofs of the theorem, see Pearl 2000; Spirtes et al. 2000; Steel 2005, 10). The causal Markov condition implies the PCC and, hence, that any probabilistic correlations among the observed fields must have a common cause that screens off the correlations.[11] Thus, if we demand that the

---

[11] This also follows from the proof sketched in Chapter 3.

causal model satisfies the Markov condition and an initial independence assumption, then C_2 is the correct model to choose.

Both C_1 and C_2 posit precise values for the initial fields, which we cannot directly measure. Alternatively, we can reconstruct our inference to the existence of a star as common cause as assuming only that the initial fields are weak (since we observe light points against the background of otherwise very weak fields) and allow that these fields can in addition contain small random fluctuations. One way to represent this in a causal model is to represent the initial fields at a point in terms of two different variables F_i and U_i, which stand for a weak initial field and a small fluctuation, respectively. The full causal model will contain both variables F_i and U_i, but we can also consider a restricted causal model C^* that excludes the fluctuation or error variables U_i. The latter model is what Spirtes, Glymour, and Shines call a *"pseudo-indeterministic model"* (Spirtes et al. 2000, 15). The model is indeterministic, because it does not include all relevant causes, but the indeterminism is only apparent, since the model can be converted into a deterministic model by including the "hidden causes" U_i. A motivation for adopting a pseudo-indeterministic model is that, as we have seen, the retarded or advanced fields associated with sources consist in the remote past or future of nothing but very small and in the limit infinitesimal correlated disturbances, which might be too weak to be measurable.

If we adopt the framework of pseudo-indeterministic models, we can appeal to likelihood reasoning to distinguish between the two models C_1^* and C_2^*.[12] The common cause hypothesis confers a much higher probability on our observations than the separate cause hypothesis:

$$P(O_1 \,\&\, O_2 / C \,\&\, weak\ initial\ fields) \gg P(O_1 \,\&\, O_2 / weak\ initial\ fields).$$

Indeed, the probability of correlated observations of light points given the assumption of weak initial fields with random fluctuations is absurdly small. Thus, to the extent that we are justified in our assumption concerning the initial fields, we should adopt a common cause model over the separate cause model.

To sum up the preceding discussion, there are inferences in physics that do not proceed from the dynamical laws together with fully specified initial or final conditions, and these inferences are paradigmatically causal inferences involving common cause reasoning. The inferences can, as we just saw, be underwritten by the assumption that the exogenous variables – the

---

[12] Eliot Sober (1984) has argued that common cause inferences should be construed as involving a comparison of likelihoods. See also Sober (2001).

incoming fields, in our example – satisfy an independence or randomness assumption. In the case of stellar observations, the extent to which we fall short of having access to the data on a full initial- or final-value surface – and, hence, the limit to our available evidence – is particularly extreme. But the very same considerations apply to "Earth-based" physical systems as well. Indeed, a moment's reflection will show that hardly ever do we have empirical access to the data on a complete final-value surface. That is, premise (1) of the argument given earlier is false: our inferences in physics do not in general follow the schema given by dynamical laws together with full initial or final conditions. Often, our inferences need, in addition to the dynamical laws, to rely on causal assumptions, since available data leave the proper choice of dynamical model radically underdetermined. Thus, we have confirmed Mill's dictum, even if not quite in the same sense as intended by him, that the law of Causation "is the main pillar of inductive science" (System of Logic, III, v, §2).

5. An explanatory asymmetry

I now want to argue that premise (3) of the argument given earlier can be rejected as well. The premise states that asymmetric causal notions could play a legitimate and substantive role in a physical conception of the world only if either they played a substantive role in explanations or inferences in addition to the purely dynamical models or their use was justified by the character of our theories' dynamical laws. I have argued that there are inferences in physics that cannot be grounded in the use of full dynamical models. I now want to argue that causal assumptions arguably can also play a legitimate explanatory role even within the context of full dynamical models. In Chapter 4 I simply posited that the retarded Green's function is privileged in that it gets the causal structure right. In what follows I will offer an argument for that claim.

It does not follow from the fact that the dynamical equations governing a system are time-symmetric that models of these equations – that is, systems satisfying the equations – are time-symmetric as well. In fact, most models of the equations (in some intuitive sense) will not be time-symmetric, and many models representing actual physical systems are characterized by a stark asymmetry between initial and final conditions: whereas initial conditions in some sense are "typical" or random, final conditions are not random and contain delicate correlations between spatially distant substates of a system. I want to illustrate what I have in mind here once again in terms of the radiation fields associated with a star.

Purely dynamically, each total field can be represented both in terms of an initial-value problem and in terms of a final-value problem. In an initial-value problem, the total field F is represented as a sum of a source-free incoming field and a so-called retarded field associated with the state of any sources in the field's *past*: $F = F_{in} + F_{ret}$. The retarded field is a field diverging from the source. A final-value problem represents the total wave as a sum of a source-free outgoing field and a so-called advanced field associated with the state of sources in the field's *future*: $F = F_{out} + F_{adv}$. The advanced field converges into the source in the future. Despite this formal symmetry, physicists routinely single out the retarded representation as the causally or physically correct field (see, e.g., Jackson 1999; Griffiths 2004; I discuss this issue in much more detail in Chapter 7). From the perspective purely of the time-reversal invariant dynamics, this may seem puzzling: both retarded and advanced representations are representations of one and the same total field. In what sense, then, can one representation be privileged as the physically correct representation? Our preceding discussion provides an answer: the retarded representation is the representation that gets the *causal* structure right. As I suggested earlier, the retarded field function can be thought of as part of a causal model $M = <C, \Theta_D>$, consisting of a causal structure C and a set of parameters Θ_D compatible with C (see Pearl 2009, 44: def. 2.2.2). The parameters Θ_D assign a function, the structural equation to each endogenous variable. In our case the structural equations are (S_{causal}): $O^i = F^i{}_{ret}(S) + f^i(F^i{}_{in})$. On the left-hand side of the equation are the effects, the observed fields $O^i = F(t_i, x_i)$, the strength of which is given in terms of their causes: the state of the source S at the earlier retarded time and initial fields F^{in}. The free incoming fields $F^{in}{}_i$ include random disturbances or background factors, U^i. As we have seen in the previous section, exactly when the U^i are distributed randomly, the causal model defined by (S_{causal}) satisfies a principle of the common cause (PCC). Despite the similarity in appearance between (S_{causal}) and the dynamical equations governing the system (S_{causal}) does *not* have an inverse $(S_{advanced})$ given by $O^i = F^i{}_{adv}(S) + f^i(F^i{}_{out})$. The reason is that the free outgoing fields, unlike the free incoming fields, do not satisfy a randomness assumption. For an advanced representation of the kinds of fields that we observe in the presence of sources to be equivalent to a retarded representation, the source-free outgoing field must satisfy $F_{out} = F_{in} + F_{ret} - F_{adv}$ and, thus, must contain "delicate" correlations, mimicking the retarded field associated with the source in the initial-value representation.

The PCC follows from an independence or randomness assumption, under the assumption that the causal structure is complete. Conversely, the initial randomness assumption can, *pace* Horwich (1987), itself be motivated by a causal representation of the phenomena at issue (see Hausman and Woodward 1999, 553; Pearl 2000, 30; also Steel 2005, 17ff. for a critical discussion).[13] That the exogenous variables are independent follows from the principle of the common cause together with the assumption of causal sufficiency, which states that exogenous variables have no common causes. This allows at least two different perspectives on the explanatory place of time-asymmetric causal structures in physics. One can adopt the view that it is precisely the fact that initial conditions are distributed randomly that makes causal reasoning and inferences possible in physics. Or one could maintain that the PCC is primary: the difference between prevailing initial and final conditions, in this second view, is explained by the fact that our world exhibits a time-asymmetric causal structure. The second view is arguably supported by our explanatory practices, which take correlations among initial states to be mysterious or miraculous but see correlations in the presence of common causes as typical.

Consider once more the contrast between a broadcast wave and "an anti-broadcast wave," to which Earman refers in the earlier quote. The highly correlated behavior of the radio signals in the two processes is precisely the kind of phenomenon that would call for an explanation in terms of localized common causes. In the case of a radio antenna broadcasting into empty space, such an explanation is readily available: the action of the antenna acting as common cause of the field disturbances can explain the strong correlations among them. Contrast this with the process of an anti-broadcast wave collapsing into the antenna. Here, too, there is an antenna located at the point on which the radio waves are centered. Yet by hypothesis, the correlations possess no *earlier* common cause and thus, by the principle of the common cause, we would not expect such a process to occur.

The field coherently converging onto the antenna presents a solution to the dynamical equations just as much as the diverging field. In fact, as Norton emphasizes, purely dynamically and considered "atemporally," the two processes are completely equivalent: in both cases there are coordinated fields that are correlated with the localized action of a source. The symmetry

---

[13] See also Pearl, who says: "note that despite its innocent appearance in associational vocabulary, the latter assumption [of initial randomness, $Cov(U_Y, U_X) = 0$,] is causal, not statistical, for it cannot be confirmed or denied from the joint distribution of observed variable, in case U's are unobservable" (2009, 704).

between the two cases is broken, however, once we embed purely dynamical models of the two processes into richer, causal structures. The diverging field then appears to be normal and entirely to be expected, since the correlations among the disturbances in the field can be explained by their common cause. The inverse process, by contrast, seems "contrived" (as Norton says; see Norton 2009, 483), "mysterious," or "improbable" (as Earman calls it), since the correlations do not have a common cause. A causal representation of the phenomenon can account for our differing explanatory practices in the two cases.

In what sense exactly is the converging wave miraculous or improbable? It follows from the dynamical equations that a source will be associated with the coordinated behavior among distant disturbances of the field. *Purely dynamically* coherently converging waves are no more improbable than diverging waves. All that the dynamics tells us is that there will be distant correlations in the field *somewhere* – be it in the past of the action of the source in the form of a converging wave, or in the future of the source as diverging wave, or perhaps a linear combination of the two.

Moreover, dynamically there is nothing especially odd about coordinated behavior of incoming *source-free* radiation. Dynamically, as we have seen, both diverging and converging waves can be represented in terms of both an initial-value problem and a final-value problem. If we represent a *converging* wave in terms of an initial-value problem, the wave will appear as incoming source-free radiation. But if we represent the same wave in terms of a final-value problem, the converging wave appears as being associated with the source. By the same token, a *diverging* wave can be represented either as outgoing source-free radiation (in a final-value problem) or as associated with the source (in an initial-value problem).

The initial randomness assumption breaks the symmetry between the different representations. And it may be that when Earman says that the absence of converging waves might be explained by an improbability of coordinated behavior in the incoming radiation, he is appealing to nothing more than the initial randomness as de facto constraint. But the quote given earlier suggests otherwise. Earman seems to suggest that there is something apparently "near-miraculous" about one class of solutions to the dynamical equations and that there is a need to explain the absence of such "near-miracles." But it is unclear what, from a strictly non-causal perspective, this miraculousness might consist of. Coherently converging anti-broadcast waves represent dynamically perfectly possible situation, which happen to be rendered unlikely by the de facto initial conditions.

Consider the following analogy: a ball is released at the top of a wedge- or roof-shaped inclined plane and rolls down on the wedge's left-hand side. As an explanation for the ball's trajectory, we can appeal to the ball's initial conditions according to which the ball was released slightly to the left of the peak of the wedge. It also is dynamically possible, given different initial conditions, for the ball to roll down the right-hand side. Even though the ball does not follow this second trajectory, since it is incompatible with the ball's actual initial conditions, there is nothing "near-miraculous" about this second dynamically possible solution – it merely is not compatible with the actual initial conditions. Similarly, absent additional causal considerations, there is nothing miraculous or "contrived" about the non-actual solutions to the inhomogeneous wave equation – they simply are solutions corresponding to non-actual initial conditions. Only within an implicitly causal picture does it make sense to think of the non-actual solutions representing converging waves as requiring the presence of a near-miracle.

Moreover, it is natural to assume that converging radiation is improbable precisely because it would require coordinated behavior in the "incoming source free radiation from different directions of space," as Earman says (507). But again a strictly non-causal view cannot support this intuition. In a strictly non-causal picture, there is no more reason to expect that incoming radiation from different directions should be uncoordinated as there is to expect that source-free outgoing radiation at spatially distant locations will be uncoordinated.

In invoking a causal explanation of why we normally do not find coherently converging waves, I am not denying the possibility of carefully setting up such waves. The way to do this is to arrange a large number of radiating objects and set them into coherent motion such that the waves diverging from each individual source combine to an overall converging wave. But notice that in this case the coordinated behavior of the wave can once again be explained by appealing to a local common cause in its past, namely, the mechanism we used to set the collection of distant sources into coherent motion. Thus, the explanatory role of the appeal to causal structures in the present case is not to prohibit certain dynamically permissible processes; rather, the causal structures serve to explain why certain dynamically possible processes are radically improbable, while their temporal inverses are utterly familiar to us.

Earman also suggests an alternative explanation for the absence of miraculous anti-broadcast waves: "a prohibition against any truly source-free incoming radiation" (Earman 2011, 507). This prohibition is usually expressed as the condition $F_{in} = 0$, the so-called Sommerfeld radiation condition. However, if understood as positing a *strict* equality, the condition is false because of the presence of the cosmic microwave background radiation. Incoming fields are generally not strictly speaking equal to zero. But, as we have seen, merely demanding that the incoming fields be approximately equal to zero is not sufficient to exclude distant small correlations among the initial fields that might at sufficiently later times result in arbitrarily strong coherently converging fields.[14] Thus, in addition to demanding that $F_{in} \approx 0$, we also have to posit an initial randomness assumption – and we are back, it seems, within the causal framework that I have outlined earlier.

Note that the preceding discussion does not depend on the fact that an antenna or a star is a macroscopic object. The inferential and explanatory structure would be exactly the same in the case of a microscopic charge, such as a single oscillating electron. The only difference is that the disturbances in the radiation field diverging from the electron as their source would be less easy to detect empirically. The point I am making is independent of the level of grain used to describe the physical systems in question, as long as an initial randomness assumption can be posited.

Indeed, the explanatory asymmetry makes its appearances even in quantum electrodynamics, as is evident from David Atkinson's (2006) discussion of the asymmetry between pure absorptions of a photon by an atom without re-emission of a photon and pure emission without absorption. Atkinson's explicit aim is to show that in quantum electrodynamics, emission and absorption phenomena are symmetric in ways in which classical models of the processes are not. Thus, he argues that it is a mistake to claim that only pure emission of a photon by an electron is possible, but pure absorption is not. For a free electron there can be *neither* pure emissions *nor* pure absorptions. By contrast, for bound electrons *both* processes are possible. So there appears to be no asymmetry. Yet Atkinson emphasizes that

> although the absorption of a photon by an atom is possible, this can be a pure process, with no re-emission, only if the photon's energy is very finely tuned to be equal to the excitation energy of one of the excited states of the atom. This is more difficult to arrange than the inverse process. (Atkinson 2006)

Atkinson here invokes the very same kind of explanatory asymmetry as Earman and others do in the case of classical radiation. In the quantum case, as in the classical case, there is a process that requires a "finely tuned"

---

[14] This point is forcefully argued by Walther Ritz (Ritz 1908a, b). See Chapter 7.

setup that is difficult to arrange, even though the temporal inverse of the process is perfectly ordinary and what is to be expected. Put differently, described from a perspective without temporal "bias," the case of pure emissions also involves a photon with an energy "very finely tuned" to the excitation energy of an excited state: "miraculously," one might be tempted to say, the energy of the emitted photon is precisely equal to the difference between the energies of the atom's excited state and its ground state. The reason why this "fine-tuning" does not strike us as odd, however, is that the atom in its excited state is the common cause of the emitted photon and the atom's de-excited state. By contrast, in the case of pure absorption there is no common cause, and the fine-tuning requires careful arrangements. Thus, even at the level of quantum electrodynamics, the explanatory and causal asymmetry is present.

6. Conclusion

In this chapter I have shown that anti-causal arguments that appeal to the causal asymmetry as a reason for denying that causal assumptions can play a role in physics fail. Indeed, I have argued that causal assumptions play an important role, even in physics, by allowing us to perform inferences from the state of a system at one time to its state at other times even in the absence of complete data on an initial- or final-value surface. Causal assumptions allow us to solve an underdetermination problem in situations where our incomplete knowledge of a system's state at a time leaves the choice of purely dynamical models radically underdetermined. Moreover, positing causal structures also allows us to account for a pervasive explanatory asymmetry that even critics of causal assumptions in physics, such as Norton and Earman, recognize. My central case study in the present chapter involved the propagation of electromagnetic waves in the presence of sources. The main point, however, is much more general: common cause reasoning allows us to make inferences from one time to another in situations where the dynamical equations governing a system on their own are silent. I will discuss the so-called radiation – or wave – asymmetry in greater detail in Chapter 7.

CHAPTER 6

Linear response theory

1. Introduction

In earlier chapters I examined a range of anti-causal arguments and argued that none of them succeed in establishing that causal notions cannot play a useful role in physics. In Chapter 5 I began to make a positive case for the claim that causal notions do in fact serve a useful purpose in modeling physical phenomena. This chapter continues the positive argument by examining one particular theoretical framework in which an asymmetric causal constraint plays a central role: linear response theory. The purpose of linear response theory is to model the response or output of a system subject to a time-dependent external field or force as input. The general framework of linear response theory is applied in a variety of different contexts in physics, which include the following: (i) modeling the response of an electrical network to an input voltage, (ii) deriving the response of a dielectric medium to an applied electric field, (iii) modeling the response of a fluid to an external force, and (iv) derivations of quantum field theoretic dispersion relations in high-energy physics. The formal framework of response theory is also closely related to models of signal processing, which we briefly discussed in Chapter 3 in the context of feedback control systems in the LHC at CERN.

A central assumption in the theory is a time-asymmetric constraint on the system's response function – an assumption that is invariably identified as causal assumption in the literature. Here are some examples of how physicists use explicitly causal language in this context: "By strict causality we mean the condition 'no output can occur before the input'" (Toll 1956, "Causality and Dispersion Relations," 1760). "We live in a universe where cause precedes effect. This is in spite of the fact that the equations of motion, whether classical or quantum mechanical, are time reversible" (Evans and Searles 1996, "Causality, Response Theory, and the Second Laws of Thermodynamics," 5808). "Causality is a basic concept in physics – so basic, in fact, that it is hard to conceive of a useful model in which effects do not have causes" (Kinsler 2011, "How to Be Causal: Time, Spacetime, and Spectra"). The time-asymmetric condition in linear response theory is "in accord with our fundamental ideas of causality in physical phenomena" (Jackson 1975, 309) "The fundamental assumption that will engage most of our attention is known as the *causality condition*. Actually, several different conditions are known by this name. The most primitive and probably also the most intuitive one can be formulated as follows: *Primitive Causality: The effect cannot precede the cause*" (5), which is a "general physical assumption" (7) (Nussenzveig 1972, *Causality and Dispersion Relations*). "The qualitative idea of causality, involving a temporal ordering of causes and effects, is so fundamental to the physical universe that one should always try to discover its quantitative corollaries" (Pippard 1978, 112). The dielectric constant "ϵ is regular [i.e., contains no divergences – see below] in the upper-half [complex] plane is, physically, a consequence of the causality principle" (Landau 1975, 279–80). Examples such as these could be easily multiplied.

In the next section I will describe the general framework of linear response theory and the role that standard discussions in the physics literature assign to the causal assumption at its core. In Section 3 I will engage with several criticisms of philosophers of the way in which physicists tend to characterize the role of causal assumptions in the theory. Here it will be important to distinguish (i) the strong skeptical claim that, contrary to what physicists themselves appear to suggest, time-asymmetric causal principles play no legitimate and substantive role in linear theory; and (ii) the weaker claim that causal principles, albeit legitimate, are not fundamental.

2. Causality in linear response theory

The basic aim in linear response theory is to model the response of a physical system to an external influence from very general assumptions without considering the details of the interaction in question. That is, instead of specifying a detailed Lagrangian for the interaction and deriving approximate predictions with the help of a perturbation expansion, the aim is to derive putatively exact predictions from what are assumed to be general physical principles. I first will introduce the framework at the most general level before focusing on the derivation of dispersion relations in classical electrodynamics as special case.

Assume that we can represent a system's response B to an external force or an external field F by a response function $L(t_1, t_2)$. If we assume that the response is linear and time translation invariant (and hence depends only on the time difference between the force and response $t_1 - t_2$), then the response $B(t_1)$ at t_1 to a force $F(t_2)$ acting for an infinitesimal time δt_2 is given by

$$\delta B (t_1) = L (t_1 - t_2)F (t_2)\delta t_2. \tag{1}$$

Integrating (1) gives the response at t as a function of the force F at all times:

$$B (t) = \int_{-\infty}^{+\infty} L (t - \tau)F (\tau)d\tau. \tag{2}$$

This expression is non-local in time, and the response $B(t)$ depends on the external force F at all times, both *before* and *after* t. In its most general form, the response function represents the system's response $B(t)$ as depending both on "remembered" past external forces and on "anticipated" future forces. The next step in the derivation is to argue that a response that occurs before the external force is applied is "not physically sensible" (Pippard 1978, 107) and, thus, to impose the causality condition "no output before the input":

$$L (t_1 - t_2) = 0 \quad \text{for } t_1 < t_2. \tag{3}$$

This is formally equivalent to including a step function in the response function L. The time-symmetric integral in (2) then turns into the following time-asymmetric expression:

$$B (t) = \int_{-\infty}^{t} L (t - \tau) F (\tau)d\tau. \tag{4}$$

Equation (4) represents the general causal response to an external force, which can then be applied in a variety of contexts to derive more specific relations characterizing different types of system, such as the dispersion relations in classical electrodynamics or the Klein-Kubo relations in fluid dynamics.

Standard discussions of linear response theory in the physics literature generally stress the following three points: first, the time-asymmetric equation (4) or its equivalents are derived from a time-asymmetric constraint; second, this constraint is physically well-founded and quite general; and third, the constraint is motivated by – or expresses – a causal condition.

Although physicists usually do not spell out carefully what work considerations of causality do in supporting equations (4), I think one can reconstruct the causal reasoning somewhat more explicitly and more precisely as
follows.

We begin with equation (1) as general relation between input and output,
assuming only linearity, time-translation invariance, and that the response
function $L(\tau)$ is square-integrable. This equation defines a class of models
that can be divided into subclasses for different choices of the response
function $L(\tau)$, which determines how much the input force or field at
different times contributes to the output at t. We then posit that the
relation between input and output ought to be understood causally: the
output variable $B(t)$ on the left-hand side of (2) is caused by the input
variable $F(t - \tau)$, representing forces applied at times $t - \tau$, on the right-
hand side. Formally, this can be represented through the introduction
of an asymmetric relation C that relates input to output variables, such
that the value of B is given as a (causal) function of its causes F: $B(t) =
f_c(F(t - \tau))$. The result is a class of what I want to call *potential causal
models*. Potential causal models are generated by adopting a particular
causal interpretation for the set of models of certain dynamical laws. Up to
this point, interpreting the relation between F and B causally may appear
to be merely an exercise in labeling: the physical content of the theory is
captured by the dynamical assumptions, it might seem, and calling certain
variables "causes" and others "effects" does not so far add to the theory's
factual content.

But next we postulate as an additional constraint on all *causally possible*
models that an effect cannot temporally precede its causes. That is, we
assume that the causal relation is not just asymmetric, but also temporally
asymmetric. Nussenzveig calls this condition "primitive causality."

In the framework introduced earlier, there is no spatial variation. If
we introduce spatial variables as well and postulate in addition that there
can be no direct causal link between spacelike separated events, then we
arrive at the principle of relativistic causality, which James Cushing calls
the "first-signal principle" (Cushing 1990), according to which for any two
systems that are a distance l apart, an event at the location of the first system
cannot have an effect on the second system before a time $\frac{l}{c}$ has elapsed after
the occurrence of the first event. Here c is the speed of light. The first signal
principle combines two dimensions of causal notions that I distinguished
in Chapter 1, a locality constraint and a time-asymmetric constraint.

Given a causal interpretation of (2), the condition of primitive causality
can be implemented unambiguously and in a well-defined manner: for

each ordered pair $<B(t), F(t - \tau)>$ such that $F(t - \tau)$ is a cause of $B(t)$, it must be the case that $\tau \geq 0$. The set of causally possible models is a subset of the set of potential causal models, which satisfies (3) and hence (4) in addition to (2). That is, two core assumptions in the derivation are that the relation between force and response is not merely one of determination but of causation and that the causal relation is temporally asymmetric.

The response function L is a Green's function. As is evident from (1), L specifies the effect at t_1 of a short, "delta-like" force at t_2. The total response is given by integrating over the contributions from all of the individual forces. Condition (3) ensures that the Green's function is "causal" and specifies the reaction of a system to a past disturbance. We have already seen in the last chapter how the Green's function formalism can be used in situations where we cannot solve a full-fledged initial-value problem. Something similar is true in the present case. Instead of writing down an interaction Lagrangian governing the details of the system's interaction with an external force or field, linear response theory proceeds by positing only very general constraints on the interaction.

What reasons do physicists have for adopting a causal interpretation of the relation between input and output variables? As we have discussed in previous chapters, an important guide to causal structures is provided by how we can manipulate or intervene into the values of the physical quantities. And indeed, the framework of linear response theory appears to provide an ideal setting for an interventionist conception of causation: a system is represented as being subject to an external perturbation to which it reacts. The strength and timing of the external perturbation are, at least in principle, open to manipulation. For example, linear response theory can be applied to electric circuits, such as an RL-circuit consisting of a resistance R and an inductance L in series. The two quantities of interest in this case are the voltage across the series and the voltage across the resistor, but which is the input and which the output? There is no explicit spatial variation in standard mathematical representations of this problem, but we can distinguish input and output by appealing to the notion of control: we can control the variation of the voltage across the series $V_{in}(t)$ and this results in a variation in the voltage across the resistor $V_{out}(t)$, but we cannot directly control $V_{out}(t)$ and thereby affect $V_{in}(t)$. Thus, when Norton asks, "How are we to pick out the computations that correspond to real causal processes?" (Norton 2009, 484), an interventionist framework for thinking about causal relations provides a plausible answer: causes are those variables that we can directly control, whereas the effect variables represent the response of the system to our interventions.

Another application of linear response theory is to how the strain rate in a fluid affects the pressure tensor. Here, too, an interventionist interpretation of the relation between input and output suggests itself. Moving different layers of a fluid with respect to each other through an external intervention – that is, through changes in the strain – affects the pressure in the fluid. That is, an external intervention on the shearing flow in the fluid has an effect on the pressure tensor characterizing the fluid. If we assume a causal response function, then changes in the pressure tensor occur only after the strain rate is changed. By contrast, as Dennis Evans and Debra Searles show, for an anti-causal response function, the pressure tensor changes *before* the strain rate changes (Evans and Searles 1996). Interestingly, as they also show, a causal system is overwhelmingly probable to satisfy the second law of thermodynamics while an anti-causal system is overwhelmingly probable to violate it.

A third application of linear response theory is given by the derivation of dispersions relations in classical electrodynamics, which I want to discuss in greater detail. In this case the response function of a medium is the Fourier transform of the dielectric constant ϵ, relating the electric displacement $D(x, t)$ to the electric field $E(x, t)$. There are two different ways in which we can set up this problem. We can begin by positing

$$D(\mathbf{x}, \omega) = \epsilon(\omega)E(\mathbf{x}, \omega) \qquad (5)$$

as constitutive relation for D and arrive at the response equation as its Fourier transform. Or we begin again by assuming that the total output field $D(x, t)$ is a linear functional of the input field $E(x, t)$ and that the system is time translation invariant. That is, we assume the equivalent of (2):

$$D(\mathbf{x}, t) = \int_{-\infty}^{+\infty} G(\tau)E(\mathbf{x}, t - \tau)d\tau, \qquad (6)$$

where G is the response function, in accord with Jackson's convention in Jackson (1999, sec. 7.10). As before, we then impose the causality condition that $G(t)$ vanishes for $t < 0$:

$$D(\mathbf{x}, t) = \int_{0}^{+\infty} G(\tau)E(\mathbf{x}, t - \tau)d\tau. \qquad (7)$$

The dispersion relations, relating the imaginary part of the dielectric constant ϵ, which characterizes the absorptive properties of a medium, to the real part of ϵ, which characterizes its dispersive (i.e., frequency-shifting) properties, follow then as a mathematical theorem from the fact that the Fourier transform of ϵ, $G(t)$, vanishes for negative t.

Toll (1956) gives the following intuitive argument for why imposing the causality condition implies that there can be no pure selective absorber – that is, there can be no medium whose only effect on an incoming wave is to selectively absorb waves of a certain frequency – and that there has to be a constraint relating the frequency- or shape-shifting properties of a medium to its absorptive properties (see also Nussenzveig 1972). Consider an incoming wave pulse E of a finite duration that is zero at all times before the time $t = 0$. This wave can be decomposed into its Fourier components – a large number of sine and cosine waves – each of which extends from $t \rightarrow -\infty$ to $t \rightarrow +\infty$ but which destructively interfere for all times $t < 0$. If a medium were selectively to absorb a component of the incoming wave of a certain frequency, E_ω, *without* shifting the frequencies of the remaining components, then the output wave would simply be the complement of the absorbed wave, that is, $E - E_\omega$, which is not zero for all times $t < 0$. That is, if there could be absorption without dispersion, then the output wave would be non-zero even *before* the arrival of the incoming wave. The causality condition denies that this is possible. Hence, the condition implies that selective absorption must be accompanied by a shifting of the frequencies of the remaining components.

Not all authors write down (6), or its equivalent (2), before introducing an additional causal constraint. Thus, Landau and Lifshitz directly write down (7) without deriving it from (6), motivating this by saying that "the physical agency underlying [(7)] consists in the process of the establishment of the electric polarization" (Landau 1975, 279). The displacement is related to the polarization via $D = \epsilon_0 E + P$. The time-asymmetrically causal picture that Landau and Lifshitz presuppose is that an external electric field applied to a medium affects the polarization, which together with the applied field causes the displacement field. Thus, Landau and Lifshitz's gloss, which invokes the causal term "agency," can again be supported by an interventionist conception of causation: we can directly manipulate the external applied electric field E. Changes in E have an effect on the polarization density P of the medium (via their effect on the individual bound electrons in the medium – see later discussion) and therefore on the electric displacement. Because we can influence the electric displacement through manipulations of the applied external field but not vice versa, the latter is a cause of the former.

At the heart of the derivation of the dispersion relations is a mathematical theorem, Titchmarsh's theorem, which states that the following three statements imply one another:

i) $G(t) = 0$ for $t < 0$; that is, the response function contains a Heaviside step function.

ii) The condition that the dielectric constant ϵ is an analytic function in the upper half of the complex ω-plane. That is, the function contains no singularities or infinities for complex frequencies $\omega = u + iv$ with positive imaginary part v.

iii) The dispersion relations, which specify a relation between the real and imaginary parts of ϵ on the real axis.

The mathematical theorem is an equivalence statement, and hence on its own does not allow us to conclude which of the three statements is explanatorily prior, but the use to which physicists put the mathematical result is to argue that there is a general physical condition, expressed by (i), which implies a constraint on the dispersive behavior of any medium. Moreover, physicists are univocal in maintaining that (i) is implied by, or expresses a causality condition. Thus, J. D. Jackson, anticipating his discussion of the dispersion relations later in his book, says that "a priori, any connection between [the wave number] k and ω is allowed, although causality imposes some restrictions" by imposing restrictions on $\epsilon = c^2 k^2 / \omega^2$ (1975, p. 223). He stresses that (7) is "the most general spatially local, linear, and causal relation that can be written between **D** and **E** in a uniform isotropic medium. Its validity transcends any specific model of $\epsilon(\omega)$" (1975, p. 309). The dispersion relations "are of very general validity, following from little more than the assumption of the causal connection between the polarization and the electric field" (p. 311). They are "extremely useful in all areas of physics. Their widespread application stems from the very small numbers of physically well-founded assumptions necessary for their derivation" (p. 312). If we interpret the relation between input and output – between electric field and displacement vector – causally, then (7) is the formal expression of the "natural requirement that an electromagnetic fields, vanishing at the place of the atom for all time $t < 0$ and beginning to act only thereafter, cannot cause the emission of scattered waves before the time $t = 0$" (Kronig 1946 quoted in Cushing 1990, 57). It is precisely its causal interpretation that makes (7) "physically well-founded," as Jackson says.

3. Possible objections and replies

In the last two sections I quoted extensively from the discussion of linear response theory in the physics literature to make clear how widespread the view is that the theory centrally involves a time-asymmetric causal constraint. This fact is an important piece of evidence in any philosophical account of the role of causal reasoning in physics because, I take it, scientific practice, including how scientists themselves describe the theories and principles at issue, should be an important guide in arriving at a philosophical account of scientific theorizing. Thus, I am inclined to give a great deal of weight to the fact that physicists themselves take the time-asymmetric constraint to be a physically well-founded causal assumption not in need of further justification. Causal skeptics are right when they contend that the question whether causal notions play a legitimate role in scientific theorizing cannot be addressed by appealing to a priori metaphysics but has to be answered through an examination of actual scientific theorizing. But if we expect our best science to help us in discovering whether causal notions play a legitimate role in the ways we represent the world, then we also have to take seriously what scientists themselves take the content of these sciences to be.

Of course, we should not in all circumstances accept physicists as final arbiters on the correct interpretation of the content of physics, but it seems to me a prima facie problem for a view such as Norton's, which wants to deny a substantive role for causal assumptions in linear response theory, that he is forced to dismiss physicists' appeals to causal notions as signs of "an illusion": physicists, Norton is forced to maintain, "succumb to the temptation" of appealing to causal notions as foundation, since it would be "awkward" for them to admit that a causal constraint is merely "opined" (Norton 2009).

Are there good reasons to reject a causal interpretation of linear response theory, despite its initial plausibility? In this section I want to discuss a number of objections that can be raised against the view that the appeal to causal principles in linear response theory is in fact good evidence for the claim that time-asymmetric causal structures play a legitimate role in at least some domains in physics. Many of the critical arguments I will consider have been raised by Norton, both in his general work on the role of causality in physics and in his criticism of my (2009), in which I first discussed the derivations of dispersion relations as an example of causal assumptions in physics. The objections I will consider are:

(i) The use of causal language amounts to mere "labeling" and the causal condition expresses a much more innocuous claim than the causal language in which it is usually couched suggests.

(ii) The (macroscopic) causal condition is recoverable from a purely non-causal theory.

(iii) Any legitimate causal principle would have to be strictly universal. But the principle cannot be universal, because there are contexts in which effects that precede their causes are accepted as a possibility.

(iv)   Causal constraints, to the extent that they are justified at all, are purely macroscopic constraints.

3.1. The "mere label" objection. Norton has forcefully argued that when causal notions are used in physics, they are "a crude and poorly grounded imitation of more developed sciences" (2003, 2). They function as mere labels and are ultimately "dispensable" (2003, p. 8). This criticism is echoed by Christopher Hitchcock, who maintains: "There are advanced stages in the study of certain phenomena when it becomes appropriate to eliminate causal talk in favor of mathematical relationships (or other more precise characterizations)" (Hitchcock 2007, 56). Indeed, one of Hitchcock's examples of putatively imprecise use of causal language – an example that he quotes from an earlier discussion of causal notions in physics by Patrick Suppes – concerns precisely the derivation of dispersion relations (see Hitchcock 2007, 55). Finally, responding directly to my discussion of dispersion relations, Norton argues that the assumption (3) stated earlier is more innocuous than my causal reading suggests, since (in the case of derivations of the dispersion relations) it "is really only saying that, in the cases we are considering, the dielectric charges respond to incident radiation; they do not anticipate it" (Norton 2009).

But Norton's rephrasing arguably itself makes a causal claim. Equation (2) posits a general connection between input and output that is non-local in time: the output at x and t depends on the input force at x at all other times. Condition (3) imposes a further restriction on this dependence, and the question is what justification we can give for this restriction – that is, for the demand that the output be responsive only to *earlier* inputs but not to *later* inputs. If we take the notion of a *response* to be itself a causal notion, then the condition that a response occurs only after the event to which it responds simply is an expression of the causality condition – however innocuous that assumption may strike us as being. But if we assume that the notion of "response" is strictly non-causal and marks merely a temporal distinction – responses occur *after* the external input, whereas anticipations occur *before* – then Norton is proposing an informal rephrasing of the formal constraint (3) without providing any reasons for why we should accept it. Thus, Norton's formulation either already contains a causal justification for the time-asymmetric dependence of the response field on the input field, or it merely restates the time-asymmetric dependence without offering any reason for its acceptance.

Norton is right, of course, in maintaining that the causal claim in question is a rather "thin" claim and does no more than introduce a temporal asymmetry: a system *responds* to an external force rather than *anticipates* it. That is, the only property of a causal relation that does any work in linear response theory is the asymmetry of the causal relation. And Norton's dismissal of the constraint as "only" stating this asymmetry is easy to understand as well: the asymmetry is so pervasive that it is easy for it go unnoticed or for us to downplay its significance. Yet, the thinness of the notion of cause not withstanding, the causal asymmetry plays an important explanatory and inferential role, as I have argued in Chapter 5.

3.2. The (macroscopic) causal condition is recoverable from a non-causal theory. In his reply to my (2009), Norton claims that in the case of the classical dispersion relations the causal condition can be derived from non-causal assumptions. At least for certain special cases, such a derivation can be found in standard textbooks, according to Norton. Jackson (1999), Norton claims, deduces "the condition [(3)] from standard electrodynamics for a special case (Section 7.10.B) without drawing on causality conditions. He then observes (Section 7.10.C) that this outcome is 'in accord with our fundamental ideas of causality in physical phenomena'" (Norton 2009, 479). This view is echoed by Sheldon Smith, who maintains that Jackson "had previously derived that $\epsilon(\omega) - 1 = \omega_p^2/\omega_0^2 - \omega^2 - i\gamma\omega$" (Smith 2013, 135) as an expression for the dielectric constant ϵ and then shows that the Fourier transform of $\epsilon(\omega) - 1$ is a retarded Green's function. That is, according to Norton and Smith, Jackson's explicit claims to the contrary notwithstanding, the logic of the situation is this: instead of *postulating* a causality condition as independent premise, Jackson *derives* the condition from non-causal assumptions.

But this characterization of how Jackson proceeds is extremely misleading, and to see how Norton's and Smith's reading diverges from what Jackson actually does will allow us to draw several important lessons about the role of causal assumptions in linear response theory. Jackson does not derive the causal assumption simply from the time-symmetric Maxwell-Lorentz equations, applied to a special case, as Norton suggests; nor does he simply derive a general expression for the dielectric constant $\epsilon(\omega)$, as Smith appears to maintain. Rather, Jackson shows that a *particular time-asymmetric* model for the dielectric constant ϵ satisfies the causality condition. The particular model he considers in Jackson (1999, sec. 7.10B) is, as he stresses, a simple one-resonance version of an index of refraction that he derived earlier from a particular microscopic model (1999, sec. 7.5.A):

$$\epsilon(\omega) - 1 = \frac{\omega_p^2}{\omega_0^2 - \omega^2 - i\gamma\omega}, \tag{8}$$

where γ is the damping factor and ω_0 is the resonance frequency. Jackson then shows that the response function

$$G(\tau) = \frac{\omega_p^2}{2\pi} \int\limits_{-\infty}^{+\infty} \frac{e^{-i\omega\tau}}{\omega_0^2 - \omega^2 - i\gamma\omega} d\omega, \tag{9}$$

which is the Fourier transform of ϵ, has two poles in the lower-half complex ω-plane and no poles in the upper-half plane. (Poles of a complex *function* G are values for ω, for which G diverges. The lower-half complex plane consists of those values for ω for which the imaginary part v of $\omega = u + iv$ is negative.) Thus, it follows from Titchmarsh's theorem that $G(\tau)$ satisfies the causality condition: $G(\tau) = 0$ for $\tau < 0$.

Thus, Norton and Smith are correct in maintaining that Jackson shows that a particular model of the dielectric constant satisfies the causality condition. But two features of this derivation are particularly important in our present context: First, the derivation begins from an explicitly time-asymmetric model. Thus, if, with Norton, we want to understand this derivation of the causality condition as proceeding "from standard electrodynamics for a special case," then we have to presuppose a view of standard electrodynamics that allows time-asymmetric assumptions to be part of the theory. And second, as Norton himself says, the derivation Jackson offers is only for a special case. Thus, after showing how a simple model of the dielectric constant satisfies the causality conditions, Jackson writes down the general time-asymmetric relation (7) and stresses that its "validity transcends any specific model of $\epsilon(\omega)$" (Jackson 1999, 332). And as a point of logic, a relation that is more general than any specific model cannot be derived from the particular model alone.

Norton and Smith maintain that "a principle of causality is not needed to complete dispersion theory" (Norton 2009, 480), since once we have specified a particular model for the dielectric constant ϵ the causality condition provides no *additional* constraint to what is already implied by that particular model. The idea appears to be that we simply write down a list of particular physically plausible models for ϵ and then derive for each individual case that the model satisfies the dispersion relations and condition (3). On this view, the causality condition (3) functions simply as a "shorthand" for some rather complicated physics embodied in each specific model for ϵ. There is no need, thus, for "invoking causality as restriction on the model" (Smith 2013, 135).

Yet alternatively one might be struck by the fact, first, that (3) allows us to unify different models of scattering interactions and, second, that a derivation of the dispersion relations that begins with (3) allows us to ignore the details of the medium in question and its detailed interaction with the field. We might then ask whether there is a common physical explanation for these facts. The causality condition offers just such an explanation: It is precisely because the causality condition constitutes a general constraint on all physically plausible models for ϵ that each such model satisfies the dispersion relations. And, as I argued earlier, this is how physicists themselves see the role of the causality condition. A general derivation of the dispersion relations from the causality condition has the advantage over any single specific model of the dielectric constant that, as Nussenzveig puts it, "the nature of the scatterer need not be specified beyond assuming that some general physical properties, including causality, are satisfied" (Nussenzveig 1972, 7). It is precisely because the causal assumption is common to all models of scattering interactions that the assumption explains or provides the "physical reason" for the dispersion relations – an explanation that tends to be obscured in derivations of the relations from any particular model for ϵ. And although it is true that once a specific causal model of the dielectric constant is assumed, we do not need to posit causality as an additional and independent assumption, the causal behavior of the model – that is, that $g(t)$ vanishes for $t < 0$ – is, according to Jackson, the most "fundamental feature" of the specific model, since it is that feature which "is in accord with our fundamental ideas of causality in physical phenomena" (Jackson 1999, 332).

Now, as Smith points out, the derivation of (at least one of the two) dispersion relations is not completely general, and "the model matters somewhat" (Smith 2013) in the derivation. In particular, what matters is whether the medium in question is conducting or not, since in the case of conducting media $\epsilon(\omega)$ has an additional pole on the real axis at $\omega = 0$ (see Jackson 1999, 332; Landau, 1975). Smith seems to take this fact to be an objection to the claim that the causality condition provides a general constraint on any legitimate model $\epsilon(\omega)$. But the force of this objection is difficult to see, since the causality condition is used in deriving the dispersion relations both in the case of conductors and in the case of insulators, as in fact Jackson and Landau and Lifshitz make clear, and the only difference between the two cases is the existence of the additional pole for metals, which results in an additional term in one of the dispersion relations. Moreover, the second dispersion relation – which, Landau stresses, is the more important one, because it can be experimentally tested – is completely general and is identical for both metals and insulators. Thus, the fact that the precise form of one of the dispersion relations depends on some gross features of the model in question *in addition to the causal constraint* does not imply that the causality condition does not function as a general constraint on all models.

That general principles can play an explanatory role over and above the particular models that satisfy these principles has been much discussed in other contexts. Thus, Hendrik A. Lorentz drew a distinction between two kinds of theories or principles in physics: general principles, which "express generalized experiences," on the one hand, and theories positing mechanisms, on the other.[1] For Lorentz, theories based on general principles and theories positing mechanisms both have their own distinct advantages and disadvantages. Both theoretical approaches are scientifically legitimate: the principle approach does not play only a secondary or derivative role, and its legitimacy neither depends on, nor is it undermined by, our possession of an underlying mechanism. The advantage of appealing to general principles, according to Lorentz, is that they are versatile and apply to a wide variety of phenomena, since they abstract from and are independent of "the inner constitution of bodies." Two examples of general principles that Lorentz cites are the principle of energy conservation and the second law of thermodynamics. The causal assumption in linear response theory arguably is another example of a Lorentzian general principle: the condition is generalized from experience, and it is one of a small number of general assumptions from the conjunction of which (in the case of dielectric media) dispersion relations can be derived without making any assumptions about a medium's "inner constitution."

Even if a principle is shown to be derivable, within a certain domain or for a specific model, from an underlying mechanism theory, it does not follow that the principle is dispensable. For instance, showing that a certain phenomenon follows from general principles, independently of the details of a particular model or mechanism-theory, may make the phenomenon seem less arbitrary than an account that invokes the details of a particular

---

[1] This distinction is closely related to Einstein's later distinction between principle and constructive theories. For discussions of Lorentz's view, see (Frisch 2005b; 2011b).

model may do. Also, general principles may permit exact derivations in circumstances in which derivations from corresponding mechanism theories may have to rely on approximation techniques, such as perturbation expansions. Indeed, as we have seen, Nussenzveig maintains that it is precisely because of its generality that the causality condition constitutes the physical reason for the dispersion relations. And, as Cushing (1990) shows in his detailed examination of quantum-mechanical dispersion theory and the S-matrix program, it was the promise of deriving exact predictions for quantum-mechanical scattering interactions from general physical principles without the need to make detailed assumptions about the nature of the scatterer that attracted physicists, many of whom treated the approach as complementary to a field-theoretic program.

In earlier papers Norton contrasts a putative principle of causality with the principle of energy conservation, which he claims is, unlike causal assumptions, not merely "decorative" (2003, 3) and is a universal principle "to which all physical theories must conform" (2007, 231). But the roles of the two principles – a causal constraint such as (3) and the principle of energy conservation – are more closely analogous than Norton allows. In fact, ironically, Nussenzveig draws an explicit analogy between the causal principle and energy conservation in linear response theory, arguing that both are "broad restrictions on physical theories" (Nussenzveig 1972, 6).

Consider how Norton's criticisms of the causal constraint in linear response theory might mutatis mutandis be applied to the principle of energy conservation. Since the principle of energy-momentum conservation (restricted to the special case of electromagnetic energy) can be derived from the Maxwell-Lorentz equations, it "can and should be founded upon existing electrodynamic theory alone," one might be tempted to say. Since the principle is "already recoverable in classical electrodynamics" from the fundamental equations, it seems that "we merely end up assigning an additional adjective ['energy'] to a condition we believe on other grounds." Obviously, then, the principle "is not needed to complete" classical electrodynamics (all quotes from Norton 2009). Moreover, it is not clear that the principle holds universally, since, for example, no general principle of energy-momentum conservation can be formulated in general relativity and, as Norton says, "a sometimes principle" is no principle at all. Attributing energy conservation to a system appears to be an exercise in mere labeling.

But this is all wrong, of course. The principle of energy conservation does have a legitimate place in physics as a general, though perhaps defeasible, constraint on our theories. That a theory satisfies a principle of energy conservation is treated as a theoretical desideratum, and the fact that the theory allows us to formulate such a principle increases our confidence in the theory. Similarly, that any specific model of the dielectric constant satisfies the time-asymmetric condition (3) can be understood as a desideratum on all models precisely because (3) is interpreted as a general causal constraint.

3.3. Causal constraints are purely macroscopic. The aim of linear response theory is to derive certain properties of a system from general principles, abstracting from the details of the dynamics governing the system. Thus, the theory proceeds from a "coarse-grained" macro description of the system. But at least some influential skeptics of causal notions in physics agree that time-asymmetric causal assumptions play a role in macrophysics and only deny that microphysics exhibits a causal asymmetry. For example, Huw Price (1997) argues that although there are macroscopic asymmetries that might be understood as causal, there is no asymmetry on the micro level that could support a time-asymmetric causal conception.

Price points to a close connection between causal representations and an assumption of probabilistic independence – a connection we have already encountered in previous chapters. Price argues that on the macro level a "Principle of the Independence of Incoming Influences (PI³)" holds, according to which incoming influences are uncorrelated. This principle can underwrite causal relations in the macro level. But according to him, we have no reason to accept a corresponding principle on the micro level, which he calls "μInnocence": "In the case of μInnocence, however, there seems to be no *observed* asymmetry to be explained," Price maintains (1997), and, therefore, there is no reason for postulating a micro principle of causation.[2]

Yet as we discussed in the last subsection, Jackson shows how the simple time-asymmetric macro model for the dielectric constant can be derived from an asymmetric microphysical model. That is, whatever asymmetry exists on the macro level is derived from a corresponding asymmetry on the micro level. The micro model assumes, as is standard in semiclassical models of atomic absorption, that electrons of the medium are bound by a harmonic restoring force and are subject to a damping force γ. That is, the bound electrons are modeled as damped harmonic oscillators:

$$m(\ddot{\mathbf{x}} + \gamma\dot{\mathbf{x}} + \omega_o^2\mathbf{x}) = -e\mathbf{E}(\mathbf{x}, t).\tag{10}$$

---

[2] As I discussed in Chapter 3, Hartry Field has a similar view.

For an external field varying harmonically as $e^{-i\omega t}$ with frequency ω, the dipole moment contributed by each electron is then given by

$$\mathbf{p} = -e\mathbf{x} = \frac{e^2}{2m}(\omega_0^2 - \omega^2 - i\gamma\omega)^{-1}\mathbf{E}. \tag{11}$$

Summing over the contributions of all the different electrons, which in general can be assumed to have different binding frequencies ω_f and damping constants γ, yields the dielectric constant. The model is semi-classical, which means (in the present case) that even though the oscillator equations are classical equations, the parameters ω_f and γ in the equation are defined through quantum-mechanical considerations.

The causality condition for the specific macro model of the dielectric constant is ultimately derived from the assumption of a time-asymmetric microscopic equation of motion for the individual electrons. Indeed, a damped harmonic oscillator subject to an external driving force itself satisfies the causality condition. The oscillator equation can be solved with the Green's function method (see Nussenzveig 1972, sec. 1.2). The Green's function $g(\tau)$ for the damped harmonic oscillator has the same analyticity properties as those of the Green's function G corresponding to our model of the dielectric constant with poles that always lie in the lower-half complex ω-plane. Hence, the model also is causal, satisfying $g(\tau) = 0$ for $\tau < 0$. As Nussenzveig (1972) says, "Since the force is considered as the cause of the displacement, this is in agreement with the *causality condition: the effect cannot precede the cause*" (14) The response is "necessarily causal, because it was derived from a causal model" (46).

Moreover, there is experimental evidence for the adequacy of the damped harmonic oscillator model for appropriately chosen values for the binding frequency and the damping constant. For example, we can investigate the absorptive properties of a medium near its resonance frequencies, where the damping coefficient dominates the dielectric constant. In metals the value of the damping constant can also be calculated from the conductivity (see Jackson 1999, 312).

That bound electrons satisfy the causality condition is related to the sign of the damping force γ. For a negative damping term, $\gamma < 0$, the Green's function would have poles in the upper-half complex plane instead of the lower-half plane and hence would satisfy an anti-causal condition instead of the causality condition, as I will explain in more detail later (see Nussenzveig 1972 14). In the classical model, the damping force γ is introduced phenomenologically and represents interactions between an electron and other particles in the material that are not represented explicitly.

Nevertheless, the model is a microscopic model of individual electrons. The interactions that are summarized in the damping force are interactions between an electron and other particles of comparable length scales and energies. The sources of the damping effect are collisions between electrons as well as lattice vibrations, lattice imperfections, and impurities (Jackson 1999, 312). These interactions result in a damping of the electron's motion, because they are random. Anti-damping would require "delicately coordinated" interactions between the electron and its surroundings conspiring to further amplify the effect of the driving force. That is, damping is associated with what Price calls an independence of incoming influences, and what I in the previous chapter called an initial independence assumption, even on the micro level. Therefore, to the extent that the damped-oscillator model for bound electrons is appropriate, we do have reasons to accept μ Innocence: we do have good reasons to accept that even microscopic particles, such as individual electrons, can be treated as independent of one another before they interact, but not after the interaction.

Two implications of this discussion are worth stressing: first, time-asymmetric causal representations play a role even on the level of microphysics, for example, in the semiclassical representation of individual electrons as damped bound oscillators. Second, once again (and in agreement with Price's view), we see a close connection between a causal principle and an assumption of initial independence. The bound electrons are damped, because their interactions with other particles in the medium are random. The oscillating electrons undergo radiation damping: in the absence of "carefully set up" interactions with other particles and the surrounding field that supply energy to the atom, the oscillating electron radiates energy. This follows from the retarded solution to the wave equation for boundary conditions that posits that no coherent field focused on an individual oscillator is coming in from past infinity. Conservation of energy then implies damping. Thus, an initial independence assumption plays a role even in the domain of microphysics. Moreover, this assumption is not restricted to the context of classical theories and also plays a central role in the discussion of damped quantum oscillators (see, e.g., Glauber and Man'ko 1984). In any such case we can represent the relation between uncorrelated inputs and outputs in terms of causal structures, where the causal Green's function specifies the structural equations linking the causal inputs to the outputs.

Before continuing my discussion of objections to my presentation of the role of causal assumptions in linear response theory, I want to pause briefly to discuss the harmonic oscillator model in a bit more formal detail. This will enable me to make explicit the differences between the Green's function formalisms in the two central case studies examined in this book: the time-symmetric wave equation used to model radiation phenomena, which we encountered in Chapter 5 and will examine in more detail in Chapter 7, and time-asymmetric linear response theory, which is the focus of the present chapter.

We have already observed the central role played by Green's functions in representing physical phenomena causally. In the context of linear response theory, there is a unique Green's function that gives the response of a system to a short, delta-function external disturbance. By contrast, in the case of undamped wave and radiation phenomena, the representation is not unique, and additional arguments are needed to establish that the "causal" Green's function is in some sense privileged. Many of the relevant differences between the two examples can be illustrated by contrasting the case of a time-asymmetric damped harmonic oscillator with that of a time-symmetric undamped oscillator. (For a more detailed discussion see, e.g., Nussenzveig 1972. The two oscillator models are also discussed in Smith 2013.)

3.4. The Green's function(s) for the harmonic oscillator. The equation of motion for an undamped harmonic oscillator is

$$\ddot{x}(t) + \omega_o^2 x(t) = f(t). \tag{12}$$

Here ω_o is the resonance frequency of the oscillator and $f(t)$ the external driving force. If we want to solve this equation with the help of the Green's function method, we have to replace the inhomogeneous source term $f(t)$ with a short, instantaneous force pulse, represented by the Dirac delta function. The Green's function $g(t)$ for the simple harmonic oscillator, thus, satisfies the following equation:

$$\ddot{g}(t) + \omega_o^2 g(t) = \delta(t). \tag{13}$$

The solution for an arbitrary function f can then be obtained by integrating over the responses to a collection of pointlike pulses:

$$x(t) = \int_{-\infty}^{\infty} g(t - t') f(t') dt'. \tag{14}$$

The Green's function can be computed using the method of Fourier transforms. The Fourier transform $F(\omega)$ of a function $f(t)$ is defined as follows:

$$F(\omega) = \frac{1}{\sqrt{2\pi}} \int\limits_{-\infty}^{\infty} f(t) e^{i\omega t} dt. \tag{15}$$

$F(\omega)$ is the frequency representation of the function $f(t)$. Using (15), we can derive from (13) an equation for the Fourier transform $G(\omega)$ of the Green's function $g(t)$:

$$(-i\omega)^2 G(\omega) + \omega_0^2 G(\omega) = \frac{1}{\sqrt{2\pi}}, \tag{16}$$

where we have used the fact that the Fourier transform of a derivative dg/dt is $(-i\omega)$ times the Fourier transform of $g(t)$ and that the Fourier transform of the delta function, up to the factor of $\frac{1}{\sqrt{2\pi}}$, is 1:

$$\frac{1}{\sqrt{2\pi}} \int\limits_{-\infty}^{\infty} \delta(t) e^{i\omega t} dt = \frac{1}{\sqrt{2\pi}}. \tag{17}$$

Thus,

$$G(\omega) = \frac{1}{\sqrt{2\pi}} \frac{1}{\omega_0^2 - \omega^2}. \tag{18}$$

Appealing to the inverse of (15), the Green's function $g(t)$ in the time representation becomes

$$g(t) = \frac{1}{2\pi} \int\limits_{-\infty}^{\infty} \frac{e^{-i\omega t}}{\omega_0^2 - \omega^2} d\omega. \tag{19}$$

Now, the problem with this equation is that the function under the integral is singular (that is, is not well defined) for $\omega = \pm\omega_0$. One way to proceed at this point is to add an infinitesimal "correction" $-i\epsilon$ in the denominator:

$$g(t) = \frac{1}{2\pi} \int\limits_{-\infty}^{\infty} \frac{e^{-i\omega t}}{(\omega_0 - i\epsilon)^2 - \omega^2} d\omega. \tag{20}$$

This integral still is along the real ω-axis, but now has poles for $\omega = \omega_0 + i\epsilon$ and $\omega = -\omega_0 - i\epsilon$.

The resulting integral can be integrated via contour integration in the complex plane. For this we evaluate the integral along a semicircular contour, consisting of the real axis and an arc in either the upper- or lower-half complex plane, and let the radius of the arc go to infinity. For $t > 0$, the contour needs to be closed in the lower-half plane, since $e^{-i\omega t}$ diverges for positive imaginary ω-values tending toward infinity. For $t < 0$, the contour needs to be closed in the upper-half plane. The method of residues tells us that the value of the integral along the contour is equal to the sum of the poles enclosed by the contour. Since the integral is regular in the upper-half complex plane, the value of the integral is zero for $t < 0$. For $t > 0$, and letting ϵ go to zero, the integral is $g_{ret}(t) = 1/\omega_0 \sin \omega_0 t$. Here g_{ret} is the retarded Green's function, according to which the oscillator is at rest before the delta-function pulse and then oscillates harmonically.

Alternatively, we could add a correction $+i\epsilon$. The result would be the advanced Green's function, which is zero for $t > 0$ and for $t < 0$ is equal to $g_{adv}(t) = -1/\omega_0 \sin \omega_0 t$. The advanced Green's function seems to suggest a system that "responds" to an external force *before* the external pulse is applied. Thus, if we take the Green's function to represent the response of the oscillator to an external "kick," then it seems intuitively compelling to reject the advanced Green's function as "non-physical."

As we will see in the next chapter, the solution for a single undamped harmonic oscillator is in crucial respects similar to the case of the Maxwell equations and radiation phenomena, since one can conceive of the electromagnetic field as consisting of an infinite number of coupled harmonic oscillators. The wave equation, which can be derived from the Maxwell equations, also has both a retarded and an advanced Green's function. Nevertheless, as we will see, there are good reasons for privileging the retarded Green's function.

The relation between the Green's function method for the undamped oscillator and for the damped oscillator that plays a role in linear response theory is this. Adding the $-i\epsilon$ correction is mathematically akin to adding a damping term to the oscillator equation:

$$\ddot{x}(t) + 2\gamma \dot{x} + \omega_0^2 x(t) = f(t). \tag{21}$$

The integral for the Green's function in this case has poles only in the lower-half plane, and $g(t) = 0$ for $t < 0$ and $g(t) = e^{-\gamma t}[\sin((\omega_0^2 - \gamma^2)^{1/2} t)]/(\omega_0^2 - \gamma^2)^{1/2}$ for $t > 0$. In the case of an *undamped* oscillator there is, at least on first sight, no obvious reason to choose a particular sign for the ϵ-correction, since the correction will be taken to be zero in the limit, and hence the Green's function corresponding to the equation is not unique. In this case causal considerations such as the ones discussed in Chapter 5 and to be discussed in Chapter 7 provide an additional constraint. In the case of the *damped* oscillator, by contrast, the sign of the damping term – that is, the fact that the oscillator is damped rather than anti-damped – ensures that there is a unique (and causal) Green's function.

The time-asymmetry in the case of linear response theory can also be seen as follows:

If we begin with (5) and, in analogy with a final-value problem for the wave equation, try to calculate the input field in terms of the output field, we can write:

$$\mathbf{E}(\mathbf{x}, \omega) = \epsilon(\omega)^{-1}\mathbf{D}(\mathbf{x}, \omega). \tag{22}$$

Equation (22) is in the frequency representation of the fields. If we wanted to transform this equation into one in the time representation, we would, according to the so-called Faltung theorem for Fourier integrals, arrive at the following equation involving the Fourier transforms of the fields (in analogy to (6) above):

$$\mathbf{E}(\mathbf{x}, t) = \int\limits_{-\infty}^{+\infty} G'(\tau)\mathbf{D}(\mathbf{x}, t - \tau)d\tau. \tag{23}$$

The problem, however, is that the Fourier transform of $\epsilon(\omega)^{-1}$ will generally not be well defined. If we assume the same simple model for the dielectric constant as before, then $\epsilon(\omega)^{-1}$ is of the form $\omega_0^2 - \omega^2 - i\gamma\omega$ and, hence, if we tried to take the Fourier transform, we would get something like

$$G'(\tau) \sim \int\limits_{-\infty}^{+\infty} e^{-i\omega\tau}(\omega_0^2 - \omega^2 - i\gamma\omega)d\omega. \tag{24}$$

In the corresponding equation (9) $G(t)$ can be evaluated by contour integration. But (24) diverges at infinity for real values of ω. More generally, if $\epsilon(\omega)^{-1}$ diverges as $\mathrm{Re}(\omega) \to \infty$, the function does not satisfy various integrability conditions that could ensure that its Fourier transform $G'(t)$ is well defined.[3]

3.5. Unless the causal principle is universal, it cannot be said to hold at all. How general is the causal assumption made in linear response theory? Does it extend to other contexts as well? Norton argues that there cannot

---

[3] Thus, Norton's identification of the two cases – that of the Green's function in dispersion theory and the Green's functions associated with the wave equation – is problematic (see Norton 2009, 481–3).

be a universal time-asymmetric causal principle, since, as he says, "there are many cases in which the effect preceding the cause is accepted as a possibility." But it is important not to conflate different senses of possibility here. It may well be that backward causation and closed causal loops are *conceptually* or *logically* possible. That is, it may well be that the forward-directedness of causation is not a necessary part of every conception of causation, but this does not imply that backward causation is *physically* possible or is possible in a universe like ours. What is physically possible is a proper subset of what is conceptually possible. Thus, it may be that positing a time-asymmetric notion of causation to represent physical processes in our universe is justified, even though the notion of backward causation is not conceptually incoherent. Exploring the properties of causal structures proceeds in ways familiar from other theoretical structures: even though in most or even all actual applications it is appropriate to assume that the causal relation is asymmetric and transitive, it may nevertheless be of interest to explore structures in which the conditions on causal relations are relaxed and which allow for causal loops. The sense in which backward causation is accepted as a possibility is perfectly compatible with the existence of an additional time-asymmetric constraint on what kind of causal models adequately represent physical systems in a universe like ours.

We also need to distinguish the notion of physical possibility from what is *possible-according-to-some-theory-T*. Arguably, relativity theory is permits various forms of backward causation. Special relativity is compatible with the existence of tachyons traveling faster than the speed of light. Einstein's equations of general relativity have solutions that contain closed timelike curves. Both kinds of solutions would result in backward causation. Even if we grant that these solutions would not engender causal paradoxes, it does not follow that backward causation is physically possible. For it might well be that the solution space of a theory outruns the space of what is physically possible.

Both the set P of what is physically possible and the set T of what is physical according to some well-supported theory T are subsets of the set C of what is conceptually possible. P and T in general are overlapping sets. If the theory T is well confirmed, then some of its models represent phenomena that are physically possible. But not all of T's models need to represent physical possibilities, since the theory may have models that violate some additional constraint on what is physically possible. Thus, neither the existence of backward causal solutions to some of our well-confirmed physical theories nor the mere conceptual possibility of backward causation show on their own that a time-asymmetric principle of causation is not universal.

Moreover, in order for time-asymmetric causal assumptions to play a legitimate role in physics, it need not be the case that these assumptions are thought to hold strictly and universally. The causal assumption in linear response theory seems to be treated as a deeply held but defeasible constraint, which is thought to constitute a broad restriction on physical theories but nevertheless is open to test and may fail in some domain. Again, I take this to be similar to the role played by the principles of energy conservation or of Lorentz invariance. These principles, too, might be shown to fail in some domain. But if they did, this would not undermine their usefulness – and indeed indispensability – in other domains.

Norton worries that once it is conceded that a causal principle is not universal, its conditions of applicability become obscure. What sorts of processes, Norton asks, are properly labeled as causal? The answer is: at least any process that is adequately represented by the kind of causal model introduced in previous chapters, which have the following features: the structural equations representing effects as function of their causes are given by the causal Green's function; the direction of causation is underwritten by the direction of interventions into the system or by the asymmetry of independence in that the causal model satisfies an initial but not a final independence assumption.

3.6. Are causal principles signs of a mere illusion or are they just not fundamental? Skeptics of the role of causal notions in linear response theory might argue for a stronger or a weaker thesis. The stronger thesis asserts that causal notions can play no legitimate role in the theory. They are, in Norton's words, a sign of an "illusion" and reflect the fact that it would be awkward for physicists to admit that the time-asymmetric constraint is merely opined and not scientifically well founded. The weaker thesis asserts that although causal assumptions are a legitimate part of the framework of linear response theory, they are not fundamental. My discussion in the preceding sections shows that the stronger thesis is implausible, but the weaker thesis cannot be dismissed so easily.

The causality condition that the Green's function vanishes for $t < 0$ is mathematically equivalent to the fact that its Fourier transform $G(\omega)$ is analytic (and hence has no poles) in the upper-half complex plane. As we have seen, the physical interpretation that physicists propose for this purely mathematical result is that the causality condition constitutes the physical basis for the dispersion relations or for related relations in other domains governing the linear response of a system. By contrast, Smith suggests that one could take the presence of damping as starting point and then derive the causality condition from this. The existence of damping is closely related to an assumption of initial randomness. Thus, one might hold that both the fact that systems exhibit damping and the fact that they satisfy the causality condition are a consequence of the more fundamental fact that initial states are distributed randomly. Alternatively, one could posit the causality principle as basic and argue that it is precisely the fact that initial states do not have common causes in their past that accounts for the randomness assumption and, hence, for the presence of damping rather than anti-damping.

Which of these two views is correct? A prior question is: should we expect this question to have an unequivocal answer? A negative answer to the latter question is suggested by Richard Feynman's Babylonian conception of physics, which arguably fits the overall practice of physics even better than a commitment to either of the two starting points as truly fundamental.

According to Feynman, a theory's theorems provide us with an inter-connected and overconnected structure that allows no unique and context-independent way of singling out certain of its parts as the most fundamental. Thus, he says, we could "start with some particular ideas which are chosen by some kind of convention to be axioms" (Feynman 1965, *The Character of Physical Law*, p. 47), but we could have chosen a different starting place as well. By contrast with what he calls "the Euclidean conception," the Babylonian conception of science is non-hierarchical and less tightly organized: "I happen to know this and I happen to know that, and maybe I happen to know that, and I work everything out from there. Tomorrow I may forget that this is true, but remember that something else is true, so I can reconstruct it all again. I am never quite sure of where I am supposed to begin or where I am supposed to end. I just remember enough all the time so that as the memory fades and some of the pieces fall out I can put the thing back together again every day" (*ibid.*). What Feynman here picturesquely describes as being a consequence of his unreliable memory is the view that there is no uniquely correct way of axiomatizing (in some loose sense) a theoretical framework. Which principles we pick as starting points is a matter of convention and may depend on the particular context in which we are applying the theory. That is, there simply may not be a uniquely correct answer to the question which of the different starting points is, as Feynman says, "more important, more basic."

Feynman's own illustration of the Babylonian conception is the relation between Newton's three laws together with the law of gravity, on the one hand, and Kepler's second law and angular momentum conservation, on the other. As Newton himself showed, we can derive Kepler's law that equal areas are swept out by a planet in equal times from Newton's laws, which might suggest that the latter are more fundamental than the former. However, Kepler's area law can be thought of as a special case of the principle of angular momentum conservation, and the latter applies much more broadly than just to gravitational forces. In fact, according to Noether's theorem, which establishes a connection between symmetries and conserved quantities, angular momentum conservation follows from the fact that the Lagrangian of a system is rotationally symmetric. Thus, one might think that the symmetry principle and angular momentum conservation together are the more fundamental principle and should be an axiom instead of the gravitational law. Yet although this principle provides a constraint on the possible form forces between bodies might take, it does not entail Newton's inverse-square law and, hence, we cannot make do with the symmetry principle alone.

Applied to the present case, it is to some extent arbitrary, and may depend on the context, whether we posit a time-asymmetric causal principle as fundamental or whether we choose a randomness or damping assumption as starting point. There simply may not be a context-independent and uniquely correct answer to the question whether an initial randomness assumption or a time-asymmetric causal constraint is more fundamental. Moreover, we could even take yet another assumption, that of a so-called passive system, as our starting point. A passive system is a system that can absorb but not generate any energy. As Nussenzveig shows (1972, pp. 391–2), any linear passive system satisfies the causality condition. Thus, formally, multiple different starting points are possible to derive relations such as the dispersion relations in classical electrodynamics or the Klein-Kubo relations in fluid dynamics. Physicists tend to single out a time-asymmetric causality condition as playing an especially central role in these derivations. But no matter which starting point one chooses, the causality condition plays an important role in the relevant structure and cannot be dismissed as scientifically any less legitimate than any of the other assumptions in the structure of theorems and principles.

4. Conclusion

Causal assumptions are especially useful in contexts in which a phenomenon cannot (or can only be with difficulty) be represented by a full initial- or final-value problem. In the previous chapter I argued that appeals to causal structures can help in solving an underdetermination problem, if we only possess incomplete data on a time slice. In the present chapter we have seen how representing the interaction of a system with an external force or field in terms of a memory or response function can enable us to investigate certain properties of the system's response without knowledge of the detailed dynamics governing the system. In both types of cases, causal Green's functions play an important role, which can be understood as defining the structural equations in a causal model. And in both cases, the causal models satisfy an initial independence assumption. In the case of a time-symmetric equation, like the wave equation discussed in the next chapter, the initial independence assumption provides a reason for privileging the retarded Green's function over the advanced Green's function. In the case of linear response theory, an assumption of microscopic independence – what Price calls "μInnocence" – can support our adoption of a time-asymmetric dependence between input and output in terms of retarded Green's functions, whose inverses are not well defined.

CHAPTER 7

The radiation asymmetry

1. Introduction

In Chapter 5 we already encountered the fact that waves and radiation fields exhibit a temporal asymmetry in the presence of wave sources. In this chapter I will examine this asymmetry in greater detail. When electric charges accelerate coherently, for example, in an antenna, we observe a radiation field coherently diverging from the source. The time-reversed phenomenon – that is, radiation waves coherently converging into an accelerating source – is not something we observe. A similar asymmetry characterizes microscopic radiation processes, for example, when a charged microscopic particle passes through matter and as a result of collisions undergoes acceleration (or rather de-acceleration). The radiation emitted during atomic collisions is called *Bremsstrahlung*, or "braking radiation" (see Jackson 1999, Chapters 13–15). The time-reverses of these phenomena, involving coherently converging waves, do not seem to exist in nature – neither on the microscopic level, concerning a small number of charges, nor on the macroscopic level, when coherently accelerating macroscopic collections of charged particles are involved.

What makes this asymmetry appear especially puzzling is that the fundamental equations governing classical radiation phenomena – both the microscopic and the macroscopic inhomogeneous Maxwell equations and the wave equation that can be derived from them – are time symmetric. Why, one might ask, do we observe only coherently diverging waves, even though the underlying laws of nature permit both diverging and converging waves? Now, most solutions to a set of time-symmetric equations will *not* be time-symmetric. Nevertheless, it seems striking that all the phenomena we observe exhibit *the same* time-asymmetry.

In discussing the asymmetry, most physics textbooks appeal to what they explicitly characterize as a causal asymmetry. Thus, David Griffiths, in his widely used textbook on classical electrodynamics, describes the asymmetry as follows (in a passage from which I quoted earlier):

> Although the advanced potentials [i.e., converging waves – see later discussion] are entirely consistent with Maxwell's equations, they violate the most sacred tenet in all of physics: the principle of **causality**. They suggest that the potentials *now* depend on what the charge and the current distribution *will* be at some time in the future – the effect, in other words, precedes the cause. Although the advanced potentials are of some theoretical interest, they have no direct physical significance. (Griffiths 2004, 425, emphases in original)

In an accompanying footnote, Griffiths explains: "Time asymmetry is introduced when we select the retarded potentials [i.e., diverging waves] in preference to the advanced ones, reflecting the (not unreasonable!) belief that electromagnetic influences propagate forward, not backward, in time" (*ibid.*).

Earman, by contrast, ridicules any appeal to causal considerations in this context:

> In physics [to gesture to wantabe laws by using suggestive but imprecise terminology – e.g., X produces (causes, contributes, . . .) Y] is not an acceptable practice. A putative fundamental law of physics must be stated as a mathematical relation without the use of escape clauses or words that require a PhD in philosophy to apply (and two other PhDs to referee the application, and a third referee to break the tie of the inevitable disagreement of the first two). (Earman 2011, 493–4)

Earman's discussion of the radiation asymmetry, like many others in the philosophy literature, includes a reference to what may be the earliest debate of this issue in the literature: a debate between the physicists Walther Ritz and Albert Einstein in the years 1908 and 1909, carried out in a series of papers in the journal *Physikalische Zeitschrift*. The debate concluded with a famous and oft-cited joint letter in which Ritz and Einstein summarized their opposing views on the arrow of radiation (Ritz and Einstein 1909). Whereas Ritz thought that the asymmetry is due to an asymmetry in the fundamental principles governing radiation phenomena, Einstein appears to have maintained that the irreversibility of radiation processes can be given a purely probabilistic explanation. Usually commentators cite the joint letter in order to appeal to Einstein's view in support of their own accounts and to argue that Einstein's view ultimately prevailed (see, for example, Norton 2009).

Discussions of the Ritz-Einstein controversy do not usually go beyond references to the joint letter. Yet once we consider this letter in its context and take into account the papers by Ritz and Einstein (Ritz 1908a; 1908b; 1909; Einstein 1909a) preceding the joint letter, as well as a paper by Einstein published later in 1909 shortly after Ritz's untimely death that year (Einstein 1909b), a considerably more nuanced picture of both Ritz's and Einstein's views emerges. Ritz, whose own theory was an action-at-a-distance theory, offered several subtle criticisms of attempts to account for the asymmetry within a field-theoretic setting that put constraints on any viable explanation of the asymmetry and point to the shortcomings of at least one standard present-day account. Moreover, Einstein's last paper on the subject in 1909 raises a vexing interpretive puzzle concerning what Einstein's view on the asymmetry of radiation in classical electrodynamics were in that year. Indeed, it appears that, contrary to the received view, it was Ritz's view that prevailed, at least in 1909, and that by the end of 1909 Ritz had convinced his former university classmate Einstein that, within classical radiation theory, the irreversibility has its source in a fundamental asymmetry of elementary radiation processes.

Thus, one of my aims in the present chapter is to set the historical record straight: Ritz's views on the arrow of radiation are far more interesting and nuanced than his being cast in the role of Einstein's foil suggests; and Einstein's views are far too ambiguous and unclear for him to comfortably play the role to which he is usually assigned – that of the "hero" coming to the defense of a non-causal account of the radiation asymmetry. After providing in Section 2 a brief introduction to a contemporary understanding of what the (or at least one) temporal arrow of radiation consists, I will in Section 3 give a critical overview over the core arguments of each of Ritz's and Einstein's papers (which originally appeared in German or French). My interests, however, are not purely historical, and it seems to me that contemporary discussions of the radiation arrow can benefit from a richer and less caricaturized understanding of the Ritz-Einstein debate. In Section 4 I will defend an appeal to causal assumptions in accounting for the asymmetry, focusing especially on Earman's forceful criticism. Among other things, I will show that the account Earman himself proposes arguably amounts to a causal account and at the very least is compatible with a causal understanding of the asymmetry. Thus, I will defend the view that causal structures play a scientifically legitimate explanatory role in representing radiation phenomena and that invoking causal notions in this contexts amounts to more than illegitimate "philosophy speak," as Earman maintains.

2. The arrow of radiation

I said that one common way of characterizing the radiation asymmetry is that it consists of the fact that there are coherently diverging but no coherently converging radiation fields in nature. Yet in one important sense this characterization is false or at least seriously misleading, since it is a mathematical fact that every radiation field can be represented both as consisting partly of diverging waves and as consisting partly of converging waves. According to our contemporary understanding, classical electrodynamics is a field theory, with a dual ontology consisting of (ultimately microscopic) charged particles and electromagnetic fields. The temporal evolution of radiation fields associated with accelerating charged particles is governed by the inhomogeneous wave equation that can be derived from the Maxwell equations. Commonly this equation is solved by setting up a modified initial-value problem. The total field in a given spacetime region Ω is then given by the contributions to the field by the sources in that region with given trajectories together with a solution to the homogeneous – that is, source-free – wave equation determined by the field values (and their derivatives) on the past boundary $\delta\Omega_i$ of the spacetime region Ω. The problem is a "modified" rather than a "pure" initial-value problem, since the trajectories of the sources are assumed to be given and are not themselves determined through an initial-value problem.[1] The fields contributed by the sources in this representation are wave-disturbances diverging from the sources. That is, if we represent the total field in terms of an *initial*-value problem, then the total field is represented as a combination of source-free incoming fields and fields associated with the sources, propagating away from a source in the positive time direction along forward lightcones centered on the trajectories of the sources. The latter fields, which are diverging from the source with which they are associated, are called "retarded fields."

Equally, we can choose to represent the total field in terms of a *final*-value problem. In that case the solution to the homogeneous wave equation is given by the fields on the future boundary of Ω, $\delta\Omega_f$, and the contributions of the sources are so-called advanced fields converging into the source. Advanced fields propagate in the *negative* time direction along past lightcones centered on the trajectories of the sources with which the

---

[1] One reason for this approach is that there exists no fully satisfactory way of tackling the full initial-value problem (see Frisch 2005a).

field is associated. In the forward-time direction, advanced fields converge into a source. Thus, one and the same total field F_{total} can be represented either as a combination of source-free incoming and retarded fields or as a combination of source-free outgoing and advanced fields:

$$F_{total} = F_{ret} + F_m = F_{adv} + F_{out}. \tag{1}$$

What is more, the field can also be represented as a linear combination of retarded and advanced fields together with appropriate source-free fields. (Since the latter require mixed boundary conditions, care must be taken, however, that the boundary conditions on the past and future boundary surface are consistent with each other.)

For reasons of mathematical tractability, the fields are in the context of mathematical derivations usually replaced by the electromagnetic potentials A_{ret} and A_{adv}, which, however, are unique only up to a gauge transformation, $A \rightarrow A + \nabla \Lambda$, where $\nabla \Lambda$ is the gradient of some scalar function Λ. Because of this gauge freedom, it is usually assumed in classical electrodynamics that the potentials merely serve a mathematical auxiliary function and, unlike the fields, do not represent anything physically real. The equations connecting fields with potentials are

$$B = \text{rot } A, \tag{2}$$

$$E = -\text{grad } \varphi - \frac{\partial A}{\partial t}. \tag{3}$$

The standard solutions for the potentials of point charges, the Liénard-Wiechert potentials $A^{\alpha} = (\Phi, \mathbf{A})$, are

$$A^{\alpha}(\mathbf{x}, t) = \frac{1}{c} \int \frac{\left[J^{\alpha}(\mathbf{x}', t) \right]_{\text{ret/adv}}}{R} d^3 x', \tag{4}$$

with $J^{\alpha} = (c\rho, \mathbf{J})$ the four-vector current and $\mathbf{R} = \mathbf{x} - \mathbf{x}'$. $[\ldots]_{\text{ret/adv}}$ means that the quantity is evaluated at the time $t' = t - (R/c)$ for the retarded solution and at the advanced time $t' = t + (R/c)$ for the advanced solution. Retarded potentials result in retarded fields and advanced potentials in advanced fields.

The upshot of the preceding discussion is that how a given total field is carved up into a component field associated with the sources present and a source-free field depends on the particular representation chosen: there is no unique way to carve up the total field into a source-free component and a component associated with the field sources. Thus, from a formal standpoint, neither a purely retarded nor a purely advanced field representation appears privileged. If we choose an initial-value problem, then any fields at the initial time before the sources turn on appear as source-free fields, and the sources contribute retarded fields propagating along future lightcones centered on the trajectories of the sources. If we choose a final-value problem, then any fields at the final time after the sources "turn off" appear as source-free fields, and the sources contribute advanced fields converging into the sources along past lightcones. Just as there is no unique field that is formally associated with the sources in a given problem, there is no unique source-free field: just as the question as to what component of the total field is mathematically associated with the field sources depends on our choice of initial- or final-value problem, so does the question as to what the source-free (or "background") field is. There is no more *the* source-free field, independent of a particular choice of representation, than there is *the* field *mathematically* associated with a given configuration of sources. Without specifying a particular representation, the question as to whether sources are associated with retarded or advanced radiation has no answer, but – and this is important as well – once we specify the representation, there is nothing else we need to know in order to determine whether fields are retarded or advanced: if we represent the total field in terms of an initial-value problem, then sources "contribute" retarded fields; and if we represent the field in terms of a final-value problem, then sources "contribute" advanced radiation. Moreover, whether a certain component of the field as free field or as field associated with a source also depends on the spacetime volume. In a retarded representation, sources in the past of the initial-value surface $\delta\Omega_i$ will contribute to what appears as a source-free field within Ω. If we enlarge the volume to include these sources, then the field associated with them will be retarded.

In what sense, then, is radiation asymmetric? One standard answer is that the asymmetry consists of the fact that the free incoming but not the free outgoing fields are approximately equal to zero. If incoming fields are equal to zero, then the total field can be represented as *fully* retarded, whereas if outgoing fields are appreciably different from zero, the total field cannot, of course, be represented as being fully advanced. In fact, it follows from (1) that $F_{out} = F_{ret} - F_{adv}$, if $F_{in} = 0$, and hence: $F_{total} = F_{adv} + F_{out} = F_{adv} + F_{ret} - F_{adv} = F_{ret}$. That is, the source-free outgoing field will consist of a combination of converging and diverging waves, which cancel out the advanced field associated with each source and ensure that the total field will be fully retarded. Thus, one common way to express the puzzle of the arrow of radiation is as follows: "Why does the Sommerfeld radiation condition $F_{in} = 0$ (in contrast to $F_{out} = 0$) approximately apply in most situations?" (see, e.g., Zeh 2007, 21). The Sommerfeld radiation condition is a temporal boundary condition – an initial-value condition. Thus, the asymmetry of the total radiation fields is expressed as an asymmetry concerning prevailing temporal boundary conditions and, hence, as an asymmetry between instantaneous states of the field. I will discuss the Sommerfeld condition and Sommerfeld's own description of its role in the theory later.

3. The Ritz-Einstein debate

3.1. Ritz's "Recherches Critiques sur Électrodynamique Générale." In February 1908 Ritz published a monumental, 130-page-long critical examination of the foundations of classical electrodynamics in the French journal *Annales de Chimie et de Physique* (Ritz 1908a). Ritz there develops and defends a field-free, action-at-a-distance theory of electromagnetic interactions. The fundamental variables of Ritz's theory are particle variables and, contrary to the standard theory, both potentials and fields are introduced only as mathematically useful auxiliary quantities. The direct action of one charged particle on another, in Ritz's theory, is given by the retarded potential A_{ret}, the "Lienard-Wiechert potential." Thus, Ritz's theory is fundamentally time-asymmetric, while the time-symmetric Maxwell equations for the electromagnetic field are relegated to the status of mathematically auxiliary assumptions. Source-free potentials exist in Ritz's theory only in the sense that there can be potentials associated with sources in the past of the spacetime volume under consideration. If we consider a spacetime volume large enough to include all charges, then there are no source-free potentials. By contrast, there will in general be outgoing radiation – radiation "into cold outer space" ("gegen den kalten Weltraum").

Part of Ritz's defense of his action-at-a-distance theory consists of a critique of the standard field-theoretic understanding of electromagnetic phenomena and, in particular, of the radiation asymmetry within a field theory. Ritz was writing at a time when the possibility of an electromagnetic "world picture" – that is, of a unified physics that has electromagnetism at its foundation – still appeared to be a genuine possibility. He begins his paper by pointing out that "in recent years electric and electrodynamic phenomena have acquired an ever greater importance, encompassing optics, the laws of radiation as well as innumerable molecular phenomena." Lorentz's microscopic theory of the electron, Ritz says, opened up the prospect of a novel conception of nature in which the laws of electrodynamics are fundamental. Ritz offers a broad criticism of the project of an electromagnetic worldview, arguing that the ether and field conceptions on which the Maxwell-Lorentz theory is based are deeply problematic. In particular, he offers the following criticisms:

  i)   Formally, fields can be eliminated and can be replaced by retarded direct interparticle interactions.

  ii) The field equations have an infinity of solutions that do not represent observable phenomena, including solutions representing a perpetuum mobile (in a sense I will discuss later). In order to restrict the solutions to what is observable, retarded potentials must be introduced as an additional explicitly time-asymmetric assumption, since this restriction cannot be derived from an asymmetry in initial conditions.

  iii) There is no unique notion of the local energy of the field or the ether.

  iv) Gravitational forces cannot be reduced to electromagnetic interactions, and hence the theory cannot be universal.

  v)   Action and reaction are not equal in a theory that posits absolute velocities.

  vi) The experimental evidence at the time, especially the influential experiments by Kaufmann, does not compel us to conclude that the mass of the electron is of purely electromagnetic origin.

  vii) Maxwell and Lorentz's theory presupposes an electromagnetic ether and thereby an absolute rest frame, which is incompatible with experience and needs to be replaced by a fully relative notion of space and time.

After reviewing the Maxwell-Lorentz theory, including the introduction of retarded potentials as auxiliary device, Ritz presents his criticisms of the notions of an electric and magnetic field (Part I, §2). In the tradition of Kirchhoff and Mach (see Chapter 1), Ritz suggests that the notion of force even in classical mechanics is not unproblematic. Introducing the notion of force through analogy by appeals to our tactile experience of forces is scientifically problematic. Moreover, since mechanical forces are not directly observable and can be detected only through observed displacements of material objects, the notion of force, according to Ritz, need not be part of our fundamental physical principles. Thus, even in mechanics the concept of force should be eliminated. If we take retarded interactions between charged particles as fundamental, the notion can be similarly eliminated in electrodynamics.

Some of Ritz's arguments concerning the electromagnetic worldview are merely of historical interest. For our purposes here, the core of Ritz's discussion consists of several arguments intended to show that one can posit retarded interactions as fundamental and derive from this the Maxwell-Lorentz equations as mathematically auxiliary equations, but that one cannot similarly begin with the Maxwell equations and arrive at an explanation of the asymmetry singling out the retarded potentials as privileged. The problem is that the Maxwell-Lorentz equations are time symmetric, "while the two time directions play different roles in the retarded potentials and in the elementary actions" (Ritz 1908a, 164).

Ritz explains that the inhomogeneous wave equation (which can be derived from the Maxwell-Lorentz equations) has different types of solution: (i) retarded solutions representing waves diverging from the source "which gave birth to them" (166); (ii) advanced solutions representing waves converging onto the source from past infinity; (iii) linear combinations of the two solutions, centered on wave sources; and, finally, (iv) solutions to the source-free Maxwell equations that are combinations of converging and diverging waves and that hence may be centered on points in empty space. For example, the difference between the retarded and advanced fields associated with a source, $F_{ret} - F_{adv}$, is a solution to the *source-free* Maxwell equations. Only the first type of solution, however, represents phenomena we find in nature, and hence we need an argument for rejecting the other types of solution.

A standard proposal, as we have seen, is to appeal to the condition that the fields and their derivatives are approximately equal to zero at some time $t = 0$ in the past, which is meant to restrict the solution space of the theory to approximately fully retarded fields. But Ritz argues that this proposal is problematic for the following four reasons:

i) The condition is not satisfied for situations that involve uniform translation or rotation. More generally, Ritz claims that the condition is empirically hardly ever satisfied. Among the phenomena we are interested in modeling, there are almost none in which the incoming fields are in fact equal to zero. If these phenomena nevertheless exhibit a radiation asymmetry, then the condition cannot be a necessary condition for the asymmetry to obtain.

ii) If we merely demand that the incoming fields at $t = 0$ be very weak, then this is compatible with there being converging fields of arbitrary strength at some later time. The condition that the incoming fields are approximately zero is compatible with the existence of small coherent fluctuations, which at later and later times gain in strength, collapsing into a source. Thus, demanding merely approximately zero fields is not sufficient for the observed asymmetry. Therefore, the condition has to be that incoming fields (and their derivatives) are *strictly* zero at $t = 0$, but, Ritz maintains, this "kind of hypothesis is impermissible in physics" (167). Alternatively one could prohibit converging waves by fiat and insist that any weak field that may be present cannot be converging. But this prohibition, Ritz claims, would be question-begging in the context of a field theory and would simply presuppose the very asymmetry we are interested in explaining. As Ritz points out, the problem in the case of electromagnetic waves is that such waves are not attenuated when they propagate in empty space (since the ether is posited to have zero viscosity). A similar problem does not arise for sound waves, where a condition of approximately zero waves would be sufficient to ensure an asymmetry.

iii)　The presence of solar and stellar radiation requires that the time $t = 0$ be placed beyond the limits of anything that is knowable. "But," Ritz says, "a hypothesis as fundamental as the radiation condition may not have such an impermissible character" (167).

iv)　If we place $t = 0$ at some finite time, then it follows that the fields *prior* to that moment are advanced, converging waves. This is so because we can use the hypersurface on which the fields are zero not only as an initial-value surface to calculate the fields in its future, but also as a final-value surface to calculate the fields in its past. But not only is a fully converging field contrary to our experience – it also represent a perpetuum mobile, in a certain sense. A charge associated with a purely advanced potential continuously receives energy from the field converging from past infinity, without any other material body losing energy.

Ritz's criticisms are justified, and any attempt to account for the asymmetry within a field-theoretic framework has to be sensitive to them: there are indeed almost no circumstances in which incoming fields are strictly zero. But if we only impose the condition that incoming fields are approximately equal to zero, one cannot exclude the possibility that there are very small coherent fluctuations among the fields in distant regions that result in a coherently converging field at later times. And imposing the field at any finite time $t = 0$ has the problematic consequence that fields are fully advanced at times prior to that time. That is, under that assumption there is no global asymmetry, in the sense that the fully retarded fields for positive times will have their counterparts in the fully advanced fields for negative times. The only remaining option, which Ritz rejects on epistemological or methodological grounds, is to demand that appropriate initial conditions of strictly zero fields hold in the infinite past. I will return to this point later.

Ritz also considers a further attempt, due to Lorentz, to arrive at the asymmetry from the time-symmetric wave equations. Lorentz requires that all field disturbances be associated with charged particles and that the state of the ether be fully determined by the state of charged matter in that the ether remains "completely idle" when there are no charges present. In response Ritz points out, entirely correctly, that "there is no definite sense attached to the proposition: perturbations depending only on the aether are excluded" (171). Consider equation (1) given earlier. Depending on the representation chosen (retarded or advanced or a linear combination of the two), different components of the total field will appear as free fields, "independent of the state of matter." If we adopt a retarded representation, then Lorentz's requirement suggests that we should set $F_{in} = 0$. But if, by contrast, we adopt an advanced representation, then F_{out} should be equal to zero and, as Ritz argues, the total field would be equal to $F_{ret} + F_{in} = F_{adv}$. Because it is not the case in general that $F_{ret} = F_{adv}$, the two ways of implementing Lorentz's condition yield incompatible results. The problem with Lorentz's attempt to single out the retarded representation as privileged, Ritz emphasizes, is that "*the decomposition of a wave-field is a mathematical operation that can be done in an infinite number of different ways*" (171, italics in the original).

Ritz concludes from his discussion that the only manner of accounting for the asymmetry of radiation is by adopting a priori, as he says, the retarded potentials, "which distinguish elementary actions" (171). Thus, "*it is the formula of the elementary actions, and not the system of equations involving partial derivatives, which is the exact and complete expression of Lorentz's theory* . . . [The Maxwell-Lorentz equations] *and the notion of the ether are fundamentally incapable of expressing the set of laws of the propagation of electromagnetic actions*" (172, italics in the original).

Ritz himself does not couch his theory in causal language, but the theory can readily be interpreted causally. If the motion of a charge is changed, then because of the retarded interactions, this will affect the motion of other charges in the future but not in the past. Influences from one charge on another propagate along future lightcones. Thus, intervening into the motion of a charge is a means of affecting the future state of motion of other charged particles, but not a means of affecting their past motion. Therefore, Ritz's theory unambiguously underwrites time-asymmetric causal counterfactuals and interventionist counterfactuals that support a causal reading of the time-asymmetric interactions among charged particles. Ritz's elementary actions between charges can be straightforwardly understood as causal relations between charges, representable in a causal model.

3.2. Ritz's "Über die Grundlagen der Elektrodynamik und die Theorie der Schwarzen Strahlung."The main topic of Ritz's (1908b) paper in Physikalische Zeitschrift, which constitutes the opening move in the Ritz-Einstein debate, is the problem of black-body radiation. According to a classical field-theoretic treatment of black-body radiation, the power radiated by a black body diverges as higher and higher frequencies of the radiation field are included. The physicist Paul Ehrenfest coined the term "ultraviolet catastrophe" for this problem. Ritz argues that the problem is a consequence of treating the ether (or the electromagnetic field) as possessing independent degrees of freedom and that problem is avoided in a retarded action-at-a-distance theory with a finite number of particle degrees of freedom.

To the objection that the diverging modes of the cavity radiation in the field theory can, in his theory, be understood as the retarded fields associated with charges in the reflecting cavity walls, Ritz responds that the actual number of charges in the walls is finite. Although it is assumed in the derivation of black-body radiation that the walls are perfectly reflecting, this idealization is "impermissible" in the present context (see 497), precisely because this idealization presupposes an infinite number of charges.

In order to motivate a pure particle theory of electromagnetism, Ritz repeats his earlier criticisms of the Maxwell-Lorentz field theory and the condition that the fields be zero at some time t_0: the Maxwell-Lorentz equations have not only retarded solutions, but also advanced solutions and linear combinations of the two. As in the earlier paper, he writes the two solutions as follows:

$$f_1(x, y, z, t) = \frac{1}{4\pi} \int \frac{\phi(x', y', z', t - \frac{r}{c})}{r} dx' dy' dz', \qquad (5)$$

$$f_2(x, y, z, t) = \frac{1}{4\pi} \int \frac{\phi(x', y', z', t + \frac{r}{c})}{r} dx' dy' dz'. \qquad (6)$$

Here c is the speed of light, ϕ is the charge configuration, and both potentials f_1 and f_2 are assumed to vanish at infinity. The retarded solution f_1 specifies the potential at t as a function of the state of the sources at the earlier, retarded time $t - r/c$, whereas the advanced solution f_2 specifies the potential in terms of the state of the sources at the later advanced time $t + r/c$.

Ritz first elaborates on his criticism in (1908a) that fully advanced solutions are unphysical because they represent a physical object that receives energy from the infinite without any other object losing any amount of energy: "Such an object, which would be capable of continuously receiving energy from the aether in this matter, would have to be called a *perpetuum mobile* and is physically impossible" (1908b, 495). Ritz then repeats his criticisms of the condition that the field is zero at some time t_0. It is obvious from Ritz's discussion of this condition (and this will become important further in Ritz's disagreement with Einstein) that what Ritz means by the integrals f_1 and f_2 for the potentials are expressions for the *total* potential: defenders of a field theory claim that f_1 results when we demand that the fields (and their derivatives) be zero at some initial time t_0. That is, the hope of defenders of the radiation condition is that this condition can ensure that the field in the future of t_0 is approximately fully retarded and that this is both necessary and sufficient to capture the sense in which radiation phenomena are irreversible. But, Ritz argues, demanding that the field be zero at some time t_0 is problematic for several now-familiar reasons: the condition prohibits many physically possible situations such as uniform translation in which the fields are not approximately fully retarded but nevertheless exhibit the radiation asymmetry; it implies that the fields prior to t_0 are fully advanced; and if the condition holds only approximately ("what alone can be asserted," 496), then converging fields are not excluded, since, for a hyperbolic equation such as the wave equation, there could be very weak convergent fields at t_0 that become arbitrarily strong at some later time.

Thus, Ritz asserts that "the complete expression of the laws of radiation and of Maxwell's theory in general does not consist in the differential equations but in the elementary actions, which arise from the introduction of the retarded potentials into Lorentz's expression of the ponderomotive force" (1908b, 496). As in the earlier paper, Ritz concludes that once we eliminate free-field solutions and independent degrees of freedom of the ether from our theory, the ether becomes "a pure abstraction" and, in accord with our experiences, "completely banned from physics" (1908b, 502). But, Ritz continues, "thereby disappears one of the main foundations of the Maxwellian description of the phenomena in terms of partial differential equations, which no longer have any physical meaning but only have the status of mathematical intermediary constructs" (1908b, 502).

3.3. Einstein's "Zum Gegenwärtigen Stand des Strahlungsproblems." Einstein's first contribution to the debate with Ritz is an answer to Ritz's (1908b) and to papers by H. A. Lorentz and J. H. Jeans on the problem of black-body radiation. I will here focus on the first section of Einstein's paper, which contains his reply to Ritz.

In accord with the accepted view, Einstein appears to endorse the opposite explanatory relation between the Maxwell equations and the retarded potentials. Whereas for Ritz the latter are primary, Einstein maintains that the retarded potentials are "only mathematical auxiliary forms." Curiously, however, he also says the following, using Ritz's term "intermediary construct" ("mathematische Zwischenkonstruktion"): "It is surely correct that the Maxwell equations for empty space, considered on their own, say nothing ['sagen garnichts aus'], that they are only intermediary constructs; the same can, as is well known, be said of Newton's equations of motion or any other theory that needs to be supplemented by other theories to deliver a representation of a complex of phenomena" (Einstein 1909a, 185).

Einstein's concession to Ritz here is puzzling. Newton's equations of motion do not say anything about the phenomena in the sense that they need to be supplemented by a specific force law. Similarly, one might say that the source-free Maxwell equations say nothing about the motion of charged objects, unless we are also told how electromagnetic fields couple to sources and are given the Lorentz force law. But if we restrict our attention to regions of empty space, then the source-free Maxwell equations do allow us to set up an initial-value problem for a source-free volume. The equations fully determine the state of the field, given the state of the field on an appropriate boundary surface, and it is unclear what any other theory could contribute to the representation of the state of the field in those regions. The source-free Maxwell equations might not tell us everything, but they also do not tell us nothing. And once we add charge and current configurations, the inhomogeneous equations determine the field with sources. This contrasts sharply with the case of Newton's theory, which cannot represent the motion of an object on which a force is acting, unless supplemented by a concrete force law. One way in which one might try to make sense of Einstein's claim is that it might be made against the background of Lorentz's prohibition against any "truly" source-free fields. Then the source-free Maxwell equations (at least for infinite volumes) have only one physically acceptable solution – that of identically zero fields everywhere.

The Maxwell equations for empty space are straightforwardly intermediary constructs on an action-at-a-distance interpretation of the theory that denies the reality of electromagnetic fields, but Einstein offers two closely related arguments against such an interpretation (independently of the issue of time-reversibility). First, in Ritz's retarded action-at-a-distance theory the "energy principle" – the principle of energy conservation – does not hold locally. This is so because the energy radiated away by an accelerated charge is not balanced locally by an increase in the energy in the field and at best shows up (partially) at some later time as the energy increase of another charged particle with which the radiating charge interacts. Second, in a retarded action-at-a-distance theory, the instantaneous state of the system does not suffice to determine the system's time evolution. A light pulse emitted by a source, Einstein points out, is not represented in the system at times between the emission event and when the light pulse is received at a screen.

These two criticisms are surely correct. If we demand that energy conservation hold locally and that our theories satisfy the Markov property and represent the evolution of a system as depending only on the instantaneous state of a system, then a retarded action-at-a-distance theory must be rejected.

Einstein also criticizes Ritz's discussion of the role of the different solutions to the wave equation. Ritz, as we have seen, claims that (5) and (6), as well as linear combinations of the two, are different solutions to the wave equations and that the field theory has no satisfactory account of restricting these solutions to the retarded solution (5). Einstein argues that this involves an elementary error: the two integrals written down by Ritz, (5) and (6), are not different solutions representing different field configurations but rather constitute different representations of one and the same field. In the retarded representation, the field is represented as depending on the state of the sources at earlier times, whereas in the advanced representation, the field is represented as depending on the state of the sources at later times. The total field in both cases is one and the same – only the representation of the field is different. As Einstein puts it, "in the first case we calculate the electromagnetic field from the totality of the processes that create it, in the second case we calculate the field form the totality of absorption processes" (Einstein 1909a, 186).

But Einstein's argument is mistaken (*pace* Earman, who cites it approvingly[2]). As we have seen in Section 2 above, Einstein is correct in that the *total* field can be given a retarded or an advanced representation. But in

[2] "But Einstein (1909) claimed that the representation by means of retarded potentials is not more special than the representation by, say, a linear combination of retarded and advanced potentials, both being representations of the same solution" (Earman, 2011).

general these representations will involve source-free fields in addition to the fields associated with sources. Although according to (1) every field can be represented equivalently as sum of retarded and incoming source-free fields or as sum of advanced and source-free outgoing fields, it is not the case in general that $F_{ret} = F_{adv}$. But as we have seen, f_1 and f_2, the two fields written down by Ritz, are the purely retarded and advanced fields, respectively, and in general these will not be equal. Now, Einstein's claim that the field can equivalently be represented by the totality of the emission or absorption processes suggests that he assumes that all emitted radiation is eventually absorbed. Indeed, he maintains that *both* the assumption of retarded radiation that is emitted into future infinity and is never absorbed *and* the assumption of purely advanced radiation coming in from past infinity involve illegitimate and paradoxical invocations of the infinite. But the assumption of complete absorption is a substantial and controversial additional assumption that does not follow from the field-theoretic framework alone. Even then Einstein's further claim that any radiation processes *in a strictly finite space* can equivalently be represented as fully retarded or as fully advanced is not correct. The only reading under which the claim is true (given the full absorption assumption) is that there will be some finite but possibly very large volume such that the total field in that volume can be represented as fully advanced.

If every field could be represented as both fully retarded and fully advanced, it becomes puzzling why we take radiation fields to exhibit a characteristic asymmetry. Why does it seem to us that there are diverging but no coherently converging fields in nature? Einstein does not offer an explanation in this paper (1909a) but ends his discussion of Ritz's view with the following intriguing remark, which hints at the view that he also seems to express in the joint letter published later that year: "Moreover we cannot conclude from the fact that [pure absorption] processes are not observable that electromagnetic elementary processes are irreversible, just as we cannot conclude that the elementary motions of atoms are irreversible from the second law of thermodynamics" (Einstein 1909a, 186).

3.4.   Ritz's "Zum Gegenwärtigen Stand des Strahlungsproblems. (Erwiederung auf den Aufsatz von Herrn A. Einstein)." In his reply to Einstein, Ritz insists that the fully retarded and the fully advanced solutions to the wave equations do indeed represent different physical processes, rather than being different representations of one and the same total field. In general, Ritz insists, the fully retarded and the fully advanced fields associated with a source are not equal: "A retarded and advanced process cannot be made to coincide simply by reversing the sign of the time [that is, replacing t with $-t$]. Thus, we are here not faced with a different kind of calculation but with a different process" (Ritz 2009, 224). Ritz goes on to point out Einstein's mistake: a general solution to the field equations contains a surface integral that is independent of the state of the sources, our F_{in} and F_{out} above. F_{in} is a solution to the homogeneous Maxwell equations, which by the so-called Kirchhoff representation theorem can be shown to be equal to a surface integral over the past spatial and temporal boundaries. The standard explanation of the radiation asymmetry maintains that the surface integral is zero in the retarded representation, but this implies that F_{out} will in general not be equal to zero: "But the Lorentzian assumptions consists in the claim that when we use f_1 and presuppose a large space, then the surface integral vanishes, from which it follows that, if instead we use f_2 for the same process, the surface integral will in general not vanish" (Ritz 2009, 224). Ritz's point here is a point I stressed in Section 2 above: if the *total* field is fully retarded, then the free incoming field (the surface integral) is zero, while if we represent the very same total field in terms of advanced fields, the free outgoing field will not be equal to zero.

In reply to Einstein's objection concerning local energy conservation, Ritz argues that to the extent that solutions to the Maxwell equations represent physically observable processes, what we can derive from the instantaneous state of the field will agree with what can be derived from the integral over the retarded sources. To the extent, then, that the two formulations are observationally equivalent, the field representation cannot be superior. But the field representation also has unphysical solutions, which can be excluded only by assuming retarded interactions. Ritz concludes that until the asymmetry can be derived successfully with the help of suitable auxiliary assumptions within the field-theoretic framework, he "will view the fact that the retarded forces are the only true integrals of these equations (into cold outer space), and that in great distances energy always flows outward or at least never inward, as the root of irreversibility and of the second law [of thermodynamics]" (Ritz 1909, 225).

3.5. Ritz and Einstein's "Zum Gegenwärtigen Stand des Strahlungsproblems." Ritz and Einstein's famous joint letter constitutes the final episodes in their debate concerning the arrow of radiation. The letter's explicit aim is "to clear up the disagreement in opinion" between Ritz and Einstein, but in the end they agree to disagree, trying to make explicit the presuppositions of their respective views. There is much about the short letter that is deeply confusing and, perhaps, also deeply confused.

The letter says that "in the special cases in which an electric and magnetic process remains restricted to a finite space, the process can be represented in the form of [the integral (5)] as well as in the form of [the integral (6)] as well as in other forms." This, of course, was Einstein's claim in (1909a) – a claim that, as Ritz had correctly insisted in his earlier paper, is false, since it ignores the surface integrals, which will not in general all be zero. In particular, if a purely retarded representation is adequate, the advanced representation will, in addition to (5), in general include a free field term that is independent of the sources. Moreover, in his criticism of the standard constraint on initial conditions, Ritz had also argued that a purely retarded representation of the *fields* is not general enough and cannot adequately represent many phenomena. Thus, it is puzzling why Ritz might have now agreed with Einstein's claim that in finite volumes the total field can be given a *purely* retarded representation, as well as a *purely* advanced representation.

Einstein, the letter says, thought that it was possible to restrict oneself to considering finite spaces without restricting the generality of the discussion, whereas Ritz takes this restriction as "in principle" impermissible. The letter continues with the following oft-quoted conclusion:

> If one adopts this [Ritz's] standpoint, then experience compels us to consider the representation by means of retarded potentials as the only one possible, if one is inclined to the view that the fact of the irreversibility of radiation must already find its expression in the fundamental equations. Ritz considers the restriction to the form of retarded potentials as one of the roots of the second law [of thermodynamics], while Einstein believes that the irreversibility is exclusively due to reasons of probability. (Ritz and Einstein 1909, 324)

The view that the retarded potentials are the correct ones to use is doubly and confusingly hedged: not only does this view presuppose that we also consider radiation into the infinite but it also presupposes that we are antecedently committed to locating the asymmetry in the fundamental laws. Are we to understand this as leaving the option open that even in the case of radiation into the infinite, other representations are possible, as long as we do not attempt to locate the asymmetry in the theory's fundamental equations?

There is no reference in the letter to any of Ritz's arguments for the assumption that the asymmetry must be taken to be fundamental, or to any of his arguments against attempts to derive the asymmetry from special initial conditions. One cannot help but wonder whether this letter might not have read very differently had Ritz not been deathly ill when it was written.

The final sentence of the of the letter picks up apparently opposing suggestions by Ritz and Einstein in their earlier papers: Ritz's suggestion that the restriction of retarded potential and, more generally, the assumption that "energy flows only outward" is at the root of the second law; and Einstein's suggestion that the irreversibility of radiation processes, like the irreversibility of thermodynamic processes, ultimately has a probabilistic explanation.

3.6. Postscript: Einstein's "Über die Entwicklung unserer Anschauungen über das Wesen und die Konstitution der Strahlung." The joint letter is dated "April 1909"; Ritz died in July at age 31, having succumbed to a many-year-long fight with tuberculosis. Later that year Einstein held a talk on "the nature and constitution of radiation" that was published in *Physikalische Zeitschrift* in October of 1909. In this talk Einstein first retraces some of the developments that led to the theory of relativity and the rejection of the ether hypothesis and then examines reasons for abandoning a purely classical conception of radiation and replacing it with a quantum hypothesis.

In the talk, Einstein does not mention Ritz or their exchange, but two passages in the paper are rather remarkable in light of Einstein's criticism of Ritz's retarded emission theory of radiation. First, Einstein says that there are phenomena that indicate that "light has certain fundamental properties, which are more readily understood from the standpoint of the Newtonian emission theory of light than from the standpoint of the wave theory. Therefore I am of the opinion that the next phase in the development of theoretical physics will result in a theory of light that can be understood as a fusion of the wave- and emission theories of light" (Einstein 1909b, 817).

Second, and more important for our purposes here, he says the following about the classical wave theory of light:

> The basic property of the wave theory, which results in these problems, seems to me to be the following. While in kinetic molecular theory there exists an inverse process for every process, in which only a small number of elementary particles participate, for example for every molecular collision, this is not the case for elementary radiation processes, according to the wave theory. According to the theory familiar to us, an oscillating ion produces a spherical wave that propagates outward. The inverse process does not exist as *elementary process*. A spherical wave propagating inward is mathematically possible; but for its approximate realization an immense amount of emitting elementary structures are needed. Elementary processes of the emission of light as such are, thus, not reversible. Here, I believe, the wave theory is incorrect. (Einstein 1909b, 821, emphasis in original)

Thus, Einstein here explicitly asserts what he earlier in the very same years appears to have denied, namely, that elementary radiation processes are time asymmetric. Moreover, whereas he earlier had taken the radiation asymmetry and the thermodynamic asymmetry to have as their common root "reasons of probability," he now draws an explicit contrast between the time asymmetry of elementary radiation processes and the kinetic molecular theory. And Einstein claims that this asymmetry is characteristic of the wave theory of radiation ("Undulationstheorie"), whereas Ritz had argued that positing asymmetric elementary actions was an argument in favor of an action-at-a-distance theory. To be sure, Einstein takes the irreversibility of elementary radiation processes in the classical wave theory to be problematic. His main reason for this is that the energy of the wave is dispersed as the wave spreads from the source, which is in tension with experimental evidence suggesting that the entire emitted energy ought to be available for elementary absorption processes. Here, Einstein says "Newton's emission theory of light seems to contain more truth" (821). Nevertheless, he claims unequivocally that the classical wave theory of radiation posits irreversible elementary emission processes.

It is difficult to render Einstein's discussion here consistent with his earlier claims that the irreversibility is "exclusively due to reason of probability" and that we cannot conclude that "electromagnetic elementary processes are irreversible, just as we cannot conclude that the elementary motions of atoms are irreversible from the second law of thermodynamics." One might try to argue that when Einstein says that the irreversibility is due to reason of probability, he means that this will turn out to be the correct explanation in whatever theory ultimately proves to be adequate and that this is compatible with holding that the wave theory posits asymmetric elementary processes. But the focus of the joint letter clearly is classical radiation theory and, hence, this attempt to construe Einstein's view in a consistent manner appears strained. An arguably more plausible interpretation is that Ritz ultimately did succeed in convincing Einstein that in the classical theory, elementary radiation processes must be understood as irreversible. In his brief discussion of the Ritz-Einstein debate, Earman says that "the predominate opinion had been that Einstein prevailed" (Earman 2011, 486). At least as far as Einstein's own thinking in 1909 is concerned, this assessment appears to be wrong, and it may well be that Einstein came to agree with Ritz on the source of the irreversibility in the classical theory.

4. Critical discussion

While Ritz and Einstein's joint letter is widely cited in discussions of the radiation asymmetry (see Price 1997; Zeh 2007; Wheeler & Feynman 1945; Norton 2009; Earman 2011), the papers by Ritz and Einstein preceding the letter receive almost no attention, and Einstein's later paper is all but completely ignored. The one notable exception of which I am aware is a letter by Karl Popper to the journal *Nature* (Popper 1956b). In an earlier letter (Popper 1956a), Popper argued that the process of waves spreading on a surface of water after a stone is dropped exhibits an irreversibility that is distinct from the thermodynamic asymmetry. The reverse process of circularly converging waves, according to Popper, "cannot he regarded as a possible classical process." He went on to say that "[the reverse process] would demand a vast number of distant coherent generators of waves the coordination of which, to he explicable, would have to he shown, in [a film depicting the process], as originating from the centre. This however, raises precisely the same difficulty again, if we try to reverse the amended film" (Popper 1956a, 538). Popper's claim that a coherently converging wave would require a vast number of coherent generators is reminiscent of Einstein's claim that for a collapsing wave to be approximately realized "an immense amount of emitting elementary structures are needed." Popper himself noted the similarity in a second letter to *Nature*: "I have found since that nearly half a century ago, Einstein used a somewhat similar argument. Had I known this, I would not have written my communication" (Popper 1956b). Popper's letters to *Nature* are discussed in Price (1997), but Price misidentifies Popper's reference to Einstein, claiming that Popper is referring to Einstein (1909a) and the Ritz-Einstein debate and not to the later publication (Einstein 1909b). The divergences in Einstein's views went unnoticed.

One of the few philosophical examinations of the Ritz-Einstein debate that goes beyond a discussion of the infamous joint letter is Earman's discussion (Earman 2011), which, however, also does not mention Einstein's later paper. Earman points out that it is important to distinguish carefully between, on the one hand, the retarded and advanced fields F_{ret} and F_{adv}, which are different *solutions* of the inhomogeneous Maxwell equations, and, on the other hand, the retarded and advanced *representation* of one and the same total field F_{total}, $F_{ret} + F_{in}$ and $F_{adv} + F_{out}$, respectively. As Earman stresses, "the latter are not different solutions but merely different representations of the same solution," which, he claims, was "noted by Einstein (1909[a])." But as we saw earlier, it was Einstein who appears to have confused the distinction, when (against Ritz's claim that F_{ret} and F_{adv} are two different *solutions* of the field equations) he argued that F_{ret} and F_{adv} are two different *representations* of one and the same field. And it was Ritz who pointed out that in order to arrive at different representations of the total field, a surface integral corresponding to a solution of the homogeneous wave equations – that is, F_{in} or F_{out} – needs to be added.

Earman's main targets are what he calls "neo-Ritzian views" of the radiation asymmetry, which in agreement with Ritz's own view invoke a restriction to retarded fields yet which, unlike Ritz, propose to do so within a field-theoretic framework (see, e.g., Rohrlich 2006 and Frisch 2005a). Earman argues that, whereas Ritz's retarded action-at-a-distance view is "scientifically respectable, if not ultimately defensible," attempts to invoke a retardation condition within the context of a field theory are not "scientifically respectable" and require "chanting incantations about 'causation.'" Yet, as we have seen, the scientifically unrespectable view of positing irreversible retarded elementary emission processes in the context of a classical field theory was also Einstein's view, and the neo-Ritzian view attacked by Earman ought, thus, perhaps more appropriately to be called "neo-Einsteinian."

Earman offers the following argument for the claim that Ritz's retarded action-at-a-distance view is scientifically respectable, while appeals to an asymmetry of the elementary processes of radiation as retarded within the context of a field theory are not:

7.1   A retarded and an advanced action-at-a-distance theory make different empirical predictions.

7.2   If two theory-formulations make distinct empirical predictions, then they constitute two distinct theories.

7.3   Therefore, a retarded and an advanced action-at-a-distance theory are two distinct theories.

7.4   Therefore, positing a retarded as opposed to an advanced action-at-a-distance-theory amounts to a scientifically legitimate hypothesis.

7.5   A retarded and an advanced field "theory" of a phenomenon, by contrast, make the very same empirical predictions and posit merely two different representations of one and the same total field.

7.6   Therefore, positing a retarded as opposed to advanced *field* theory as in some sense privileged amounts to positing a difference that makes no empirical difference, and hence is not scientifically legitimate.

Here are the details of the argument. Consider a retarded action-at-a-distance theory for a collection of charges in a spacetime volume Ω, which is large enough to include all charges. Earman shows that this entails an electrodynamic arrow: If Ω includes all charges, the incoming auxiliary "field" will be zero, yet since the sum of retarded actions will not in general be equal to the sum of advanced actions, there will in many cases be an outgoing auxiliary "field" – that is, the surface integral over the future boundary of the volume will not be zero. (Recall that in an action-at-a-distance theory, the fields or potentials have only the status of auxiliary entities.)

The equation of motion for a charge states that the charge's acceleration is determined by the sum of the retarded field-forces associated with all other charges. If we contrast a retarded action-at-a-distance theory with a fully advanced theory with an equation of motion for a charge in terms of the advanced field-forces associated with all other charges, we find that the two theories make different predictions: the sum of the fully retarded potentials and the sum of the fully advanced potentials will not in general be equal, as Ritz already had argued in response to Einstein. Thus, a fully retarded and a fully advanced theory can, at least in principle, be empirically distinguished.

This is not the case, Earman argues, if one tries to single out the retarded representation as privileged within a field-theoretic framework. In this case the force on a charged particle is given by the sum of the retarded field-forces *together with the force associated with any free incoming field*. Earman insists that in this case the incoming field cannot be assumed to be equal to zero: "Allowance for the homogeneous solution must be made on pain of restricting the range of validity of the theory" (Earman 2011, 497). Similarly, an advanced field theory will include a term for a free outgoing field corresponding to a solution of the homogeneous Maxwell equations. But once we allow for the addition of arbitrary solutions to the homogeneous field equations, then *any* configurations of fields and charges can be represented in terms of *both* a retarded theory and an advanced theory. Thus, within a field-theoretic framework, the retarded and the advanced formulations are two different representations of one and the same theory, rather than two different theories. Trying to single out the retarded representations as that representation that specifies what a charge causally contributes to the field introduces a distinction without empirical content.

At this point defenders of a "retardation condition" might want to argue that prevailing initial conditions introduce an asymmetry that can single out the retarded representation as privileged. Zeh, as we have seen, expresses this asymmetry as follows: "Why does the condition $F_{in} = 0$ (in contrast to $F_{out} = 0$) approximately apply in most situations?" Thus, one could try to argue against Earman's equivalence claim that we are entitled to set the homogeneous solution approximately equal to zero in a retarded representation of the fields and thereby arrive at an asymmetry that allows to single out that representation as privileged. But Earman argues that this response fails for reasons closely analogous to ones already discussed by Ritz: "It would seem that in a natural sense of 'most' F_{in} is not approximately zero in the visible part of the electromagnetic spectrum for most of the systems of which we are aware, since otherwise we would not be aware of them. And the ubiquity of the cosmic background radiation makes one think that in a natural sense of 'most' F_{in} is not approximately zero in the microwave spectrum for most systems, whether we are aware of them or not" (Earman 2011). Hence, Earman concludes, positing a causal asymmetry can amount to no more than scientifically illegitimate "incantations of causation."

Does this argument show that invoking a fundamental time-asymmetric constraint on the level of the elementary radiation processes is scientifically respectable within the context of an action-at-a-distance theory but not within a field theory? The answer is "no," for Earman's argument sets up a false dichotomy. His discussion of action-at-distance theories takes place within the context of *idealized models* of the theory, whereas in discussing field theories he invokes the "messiness" of *actual, real-world systems*. (Thus, Earman here is guilty of a mistake analogous to the error made by Field and Woodward discussed in Chapter 3.) Once we distinguish carefully between a model and the real-world systems it is meant to represent, the apparent disanalogy between action-at-a-distance and field theories disappears. *Both* in the case of Ritz's theory *and* in the case of its field-theoretic cousin, it is often possible to represent a system in terms of a *model* that includes only a finite number of sources and contains no additional sources outside of the volume under consideration. In considering such models, it is legitimate both in the context of an action-at-a-distance theory and in the context of a field theory to posit the condition $F_{in} = 0$, without thereby illegitimately restricting the theories' range of applicability. For example, in modeling the radio signal emitted by an antenna, it may by legitimate to ignore the presence of other radiation fields, such as visible light or the cosmic background radiation, and consider *only* the fields associated with the antenna, which are strictly zero before the antenna turns on.

By contrast, when our aim us to characterize an electromagnetic asymmetry in actual systems taking into account the total actual field, then both theoretical frameworks have to allow for non-zero fields on the past boundary of any finite spatial volume. Earman's criticism of the condition of approximately zero fields applies with equal force to the action-at-a-distance theory, since fields, real or auxiliary, on any initial-value surface at a finite time will never be strictly speaking zero. In fact, both frameworks have the same representational resources at their disposal: A field theory together with the assumption $F_{in} = 0$ has the same representational resources as an action-at-a-distance theory that presupposes that there are no sources outside of the volume under considerations. Similarly, an action-at-a-distance theory that allows fictitious incoming fields associated with past sources that are not explicitly modeled has the same representational resources as a field theory without a restriction on possible initial fields.

In examining the asymmetry of radiation, it is important to distinguish carefully the following three questions:

i) What is the most precise way of characterizing the intuitively obvious empirical asymmetry exhibited by actual radiation fields?

ii) In what sense are the idealized models we use to represent radiation fields time-asymmetric?

iii) What can account for the empirical success of the time-asymmetric models we use to represent radiation phenomena? And what can explain the asymmetry of actual radiation fields?

Ritz's critical arguments, as well as Earman's discussion, point to the difficulty in finding a mathematically sharp answer to question (1) applied to finite systems in a finite spatiotemporal volume. The obvious candidate for a sharp empirical asymmetry, that incoming fields are approximately equal to zero at some initial time t_0, fails. Nevertheless, it often seems to be empirically successful to represent an actual system in terms of a model that assumes no incoming fields. That is, the answer that is inadequate as a response to (i) seems to be correct as response to (ii). But this answer, *pace* Earman, can be given by advocates of action-at-a-distance theories and of field theories alike. In response to (iii), finally, a defender of an action-at-distance theory will with Ritz invoke retarded elementary actions, while a defender of a field theory will with the Einstein of (1909b) invoke an asymmetry of elementary radiation processes. How this explanation is to work in detail is best seen by considering the case where the empirical asymmetry can be formulated most cleanly – radiation in an infinite volume encompassing all of spacetime. Thus, I am here following Earman, who maintains that a "scientifically respectable" precise electromagnetic arrow emerges as we enlarge the volume under consideration to include all of spacetime.

Before turning to the case of an infinite volume, it is worth making explicit an additional consequence of our discussion. Contrary to what both Ritz and Earman assume, an action-at-a-distance theory, too, has no immediate answer to the question as to what the observable asymmetry in the fields consists of, for in most actual cases there will be sources in the past of a spacetime volume under consideration and hence the initial fictitious fields will be different from zero. How one might capture the asymmetry in the presence of both incoming and outgoing free fields is a question I will return to later.

As a concrete example of a system radiating into infinite space, Earman considers a radio antenna in which electrons oscillate in unison. He then asks us to imagine that the antenna

> broadcasts into empty space so that the outgoing radio waves are not absorbed but travel to spatial infinity. It would seem nearly miraculous if the time reverse of this scenario were realized in the form of anti-broadcast waves coming in from spatial infinity and collapsing on the antenna. The absence of such near miracles might be explained by an improbability in the coordinated behavior of incoming source free radiation from different directions in space. Or it might be explained non-probabilistically by a prohibition against any truly source-free incoming radiation. The latter is one motivation for the Sommerfeld radiation conditions. (Earman 2011, 506–7)

The retarded Sommerfeld radiation condition is a boundary condition at infinity. Earman states this condition as the condition that the free-field component F_{in} of the total field F_{tot} vanish in the limit as the initial conditions are evaluated at past infinity. This formulation is not entirely accurate. The Sommerfeld condition is a condition on the fields *on the boundary*, rather than a condition on the solution F_{in} to the homogeneous field equations. Eventually, I want to examine Earman's explanation for the asymmetry. But first I will discuss the Sommerfeld condition in somewhat more detail, since it provides an answer to questions (i) and (ii) listed earlier and also to Ritz's worry of what a proper boundary condition may be that can ensure, within a field theory, the absence of advanced radiation.

The problem to which Ritz pointed, and which was only solved by Sommerfeld in 1912, is that the inhomogeneous wave equation (and the Helmholtz equation, which is its Fourier transform) does not have a unique solution subject only to the condition that the fields and their derivatives vanish at infinity. The constraint of vanishing fields in the limit is compatible both with a fully retarded solution and a fully advanced solution, as well as linear combinations of the two. In particular, it is also compatible with differences between retarded and advanced fields, $F_{ret} - F_{adv}$, which are standing-wave solutions to the homogeneous wave equation. These solutions represent fields that converge on an arbitrary point and then rediverge and can be added to any solution to the inhomogeneous wave equation. Thus, the boundary condition of vanishing fields at infinity leaves the solution to the wave equation radically underdetermined. This contrasts, for example, with the diffusion equation, which does have a unique solution.

There is no ambiguity in the sign of the Green's function in that case, and the condition that the function vanishes at infinity determines the Green's function uniquely.

Here is how Sommerfeld himself characterizes the problem in his paper in which he first presented the stronger boundary condition that has come to be known as the "Sommerfeld radiation condition":

> The physical reason for this is the following: In optics and similar fields, we are dealing with *propagating waves* that radiate from sources located in the finite into the infinite, i.e. *diverging waves*. Physically not realizable, but mathematically equivalent, would be waves that radiate from the infinite and disappear in source-points located within the finite, i.e. *converging waves*. By suitably combining both types of propagating waves, sources can be eliminated and we obtain *standing waves*, which have the character of eigenfunctions of the infinite domain. The possibility of superimposing such standing waves on every solution of the present problem shows the problem's lack of uniqueness. Yet, since nature, of course, instantiates a unique solution to the problem, we conclude that there must be an additional constraint, which singles out propagating diverging waves from the manifold of solutions to the wave equation. This constraint will concern the behavior of waves at infinity; we will call it *radiation condition*. (Sommerfeld 1968, italics in the original)

What Sommerfeld appears to be saying is that one problem concerning the wave equation is that its class of solutions includes solutions that are not physically reasonable. Hence, his attempt to look for a condition that can restrict the solution space of the equation to those solutions that are physically reasonable or physically possible. That is, rather than taking the wave equation as delimiting the range of what is physically possible and then looking for an explanation of why a large class of physically possible is not actualized, the problem for Sommerfeld seems to be with the mathematics: the wave equation has "too many" solutions, while nature picks a unique solution. According to this view, the Sommerfeld radiation condition does not explain the asymmetry but is merely the mathematical condition that enables us to exclude non-physical solutions of the wave equation and restrict the solutions to the physically reasonable purely diverging waves.[3]

---

[3] It appears that Hendrik Lorentz had a similar view in that he thought that the theory allowed for more solutions than are physically reasonable: "However, [the retarded potentials are] not the most general solution of the fundamental equations . . . and for example solutions are possible that show a propagation towards instead of from the volume elements. But of those we want to keep the theory free by assuming once and forever that the charged volume elements are really just starting points of disturbances of the equilibrium. We also exclude all states of the aether that do not depend on charged matter; if the latter were not there, the equilibrium of the aether would stay forever undisturbed." (Lorentz, 1904, 158–9, cited in Ritz, 1908a, 332)

In somewhat more detail, Sommerfeld arrives at his radiation condition as follows. He begins by considering the Fourier transforms of the wave equation, the *Helmholtz equation*

$$(\nabla^2 + k^2)u(x, \omega) = -4\pi\phi(x, \omega) \tag{7}$$

for a single point source. Solutions to this equation are the retarded and advanced Green's functions G_+ and G_-, as well as linear combinations of these, such as the standing wave solution:

$$G_0 \sim= \frac{1}{2i}(G_+ - G_-) = \frac{1}{4\pi}\frac{\sin kR}{R}. \tag{8}$$

The Sommerfeld radiation condition is a constraint on the behavior of u as $R \to \infty$. But since both G_0 and its derivative tend toward zero,

$$\lim_{t \to -\infty} G_0 = 0 \quad \text{and} \quad \lim_{t \to -\infty} \frac{dG_0}{dR} = 0, \tag{9}$$

requiring that the field and its derivative tend toward zero is not stringent enough to exclude incoming radiation. Sommerfeld was looking for a condition that could ensure a unique solution to the wave equation and, in particular, managed to restrict solutions to the diverging waves: "At infinity u must be representable as a sum (or integral) of waves of the divergent propagating type" (Sommerfeld 1912). As he shows, the condition is

$$\lim_{R \to \infty} R\left(\frac{dG}{dR} - ikG\right) = 0, \tag{10}$$

if we assume that the time dependence of the wave is $e^{-i\omega t}$. Intuitively, this means that the expression in parentheses involving the Green's function and its derivative has to vanish quickly enough to "make up for" the term R, which is diverging as $R \to \infty$.

Thus, Sommerfeld's condition offers an answer to (at least one way of interpreting) Ritz's or Einstein's worries about the infinite. Ritz, as we have seen, takes a constraint at a time beyond the limit of anything that is knowable to have an "impermissible character," and Einstein maintains that considering the question of emission of radiation into the infinite or from the infinite to invite illegitimate paradoxes of infinity and says: "If we want to remain within the realm of experience, then we cannot speak of the infinite but only of spaces that lie outside of the space under consideration" (Einstein 1909a, 186). Both Ritz and Einstein couch their worries in more general epistemological terms, but at least in Ritz's case part of the worry appears to have been that the constraint at infinity that fields tend to zero leaves the fields underdetermined. This worry is answered by the Sommerfeld condition: there is a mathematically precise constraint that ensures that the surface integral in the retarded field representation vanishes in the infinite limit and hence that the total fields are fully retarded.

The radiation condition provides us with answers to questions (i) and (ii) listed earlier: in the case of radiation into the infinite, there is a mathematically precise way of characterizing the asymmetry of systems involving radiating sources. But this does not yet answer question (iii) – the question as to what can account for this asymmetry. I now want to return to my discussion of Earman's answer to this question and thereby defend the view that an adequate account of asymmetry will involve an appeal to causal structures.

Earman agrees that the condition is in need of a motivation. In his discussion of the antenna broadcasting into empty space and its time-reverse, an "anti-broadcast" wave collapsing on the antenna, he says that such an anti-broadcast wave would be "near miraculous" and suggests that one explanation of the absence of such near miracles, which also provides a motivation for the Sommerfeld radiation condition, may be a "prohibition against any truly source-free incoming radiation." He proposes a second, probabilistic explanation of the absence of near miracles. I want to argue now that both proposed explanations are compatible with causal representations of the interactions between sources and fields.

First, the prohibition against any "truly source free radiation." It is important to stress that this prohibition is an explicitly time-asymmetric constraint posited in addition to the Maxwell equations. The prohibition is against source-free *incoming* radiation, and not against source-free radiation tout court, since which component of the radiation field is source-free and which is associated with field sources depends on the representation chosen. Recall (1) earlier. There will be neither source-free incoming radiation nor source-free outgoing radiation only in the special case when $F_{ret} = F_{adv}$. In general, $F_{out} \neq 0$ when $F_{in} = 0$, and $F_{in} \neq 0$ when $F_{out} = 0$.

A constraint against truly source-free incoming radiation is equivalent to a causal constraint, however, since the constraint supports what are paradigmatically causal counterfactual inferences and, in particular, inferences about counterfactual interventions into a system of charged particles. Imagine a source that is "turned on" (i.e., its charges accelerate, perhaps due to a non-electromagnetic force) and radiates for a brief period of time. We can ask what the field would have been if the source had not been turned on. The recipe for answering such questions in the case of theories that pose a well-defined initial-value problem is to take the actual state of the field on some initial-value surface as given and then use the field equations to solve an initial-value problem in which the source is turned off. But without any additional constraint, such as Earman's prohibition, the problem is underdetermined. We could take the *initial* field before the source is turned on in the actual world as given and solve an initial-value problem with no radiating source in the future of the initial-value surface. Alternatively, we could take the *final* field after the source was turned on as given and solve a final-value problem with no radiating source in the past of the final-value surface. The two answers will in general be different. If we imagine that the antenna in Earman's example emits only a brief pulse, then solving an initial-value problem tells us that if the antenna had been turned off, the total field would have been zero. But the solution to a final-value problem tells us that if the antenna had been turned off, then there would have been a wave coherently converging into the antenna and diverging from it. This can be seen by considering (1). If the free incoming field is zero in the actual world, then $F_{total} = F_{ret} = F_{adv} + F_{out}$. In a final-value representation, for the source to be turned off (in the counterfactual situation) means that $F_{adv} = 0$. Therefore, the total field in the counterfactual world is equal to F_{out}, which is the same as in the actual world and is given by $F_{out} = F_{ret} - F_{adv}$. In the case of an initial-value problem, changing the state of the source affects the field *after* the changes to the source, whereas in the case of a final-value problem, changing the state of the source affects the field *before* the counterfactual change to the state of the source. Thus, Earman's skepticism concerning the value of this kind of counterfactual reasoning may appear to be well justified. He says, "The exercise of trying to divine the truth value of such counterfactual assertions, even when it is agreed at the outset what the basic laws are, is an invitation to a contest of conflicting intuitions about cotenability of conditions and the closeness of possible worlds" (Earman 2011, 494).

The ambiguity is removed, however, by adding the prohibition against source-free incoming radiation as additional constraint. This constraint ensures that the initial-value problem gives the uniquely correct answer to how the state of the field would change if the source had been turned off, since setting up a final-value problem with the actual outgoing fields as input will imply the presence of a source-free incoming field when the source is turned off. Thus, the prohibition against source-free incoming radiation provides unambiguous truth conditions and ensures that counterfactual changes to the state of a source are associated with changes in the field only after the changes to the sources are postulated to take place.

Moreover, we can think of changes to the state of a source in interventionist terms. If we want to intervene experimentally into the state of the field, we can do this by changing the state of the source: manipulating the state of a source is a means of manipulating the electromagnetic field, and there are countless experimental applications of this connection (for example, in the LHC at CERN, which we discussed in Chapters 3 and 4). It follows from the counterfactual asymmetry underwritten by the prohibition against source-free incoming radiation that all interventions into the state of the field are from the past: we can affect the state of the field in the future but not in the past (see Chapter 4).

Experimental contexts suggest that there are good reasons for allowing such counterfactuals even within the context of a scientifically respectable discussion of the radiation asymmetry. But the counterfactual and intervention asymmetries are precisely the kind of asymmetries that are characteristic of a causal asymmetry. Thus, even though Earman himself does not want to use causal language in expressing the constraint on incoming fields, the constraint is inherently a causal constraint. It is a constraint that allows us to represent the relation between the state of a source and electromagnetic field measurements in terms of time-asymmetric causal structures that underwrite time-asymmetric causal counterfactuals and an interventionist reading of these counterfactuals. As the physicist Seth Lloyd puts it, in a paper arguing for the usefulness of applying a theory of causal graphs to physical systems, "the requirement in electrodynamics that the source-free part of the incoming electromagnetic field vanish is a way of realizing the requirement that correlated variation between the motions of charged particles be caused by the motions of charged particles in the past" (Lloyd 1996, 114).

Earman proposes a second explanation for the absence of "near miraculously" converging waves: the coordinated behavior of incoming source-free radiation from different directions in space might be radically improbable. This explanation is compatible with the existence of "truly" source-free incoming radiation and does not imply the Sommerfeld radiation condition – and, in fact, contrary to what Earman suggests, the constraint can be invoked for finite volumes as well and does not presuppose considering infinite volumes – but it, too, implies that the relationship between sources and radiation fields can be represented by causal structures, as we have already seen in Chapter 3. What Earman invokes is an initial randomness assumption. Like the prohibition against incoming source-free radiation, the constraint is inherently time-asymmetric. For it follows from the fact that incoming radiation from different directions is not coherent or coordinated that outgoing radiation in the future of a radiating source will be coordinated. The field at times after a source radiating into empty space was turned on will contain correlations, which will eventually become smaller and smaller and ever more distant – eventually leading to what viewed backward in time would look like "near miraculous" microscopic correlations among distant field regions. Of course, the distant correlations strike us as not being miraculous, precisely because they can be explained in terms of the retarded field associated with the source. That Earman finds the temporal inverse, the advanced field associated with a source, to consist of near-miraculous correlations is itself telling. But why should the coordinated behavior of *incoming* source-free radiation be radically improbable, but the coordinated behavior of *outgoing* source-free radiation be completely ordinary and to be expected?

Moreover, whether coordinated behavior of source-free incoming radiation is improbable depends crucially on whether the source with which the correlations in the field can be associated lies in the future or in the past. Consider as a slight variant of Earman's example an antenna that emits two brief radiation pulses at times t_1 and t_2. If we choose an initial-value surface at some time in between t_1 and t_2, then there will be correlations among the field vectors on the cross section of the surface with the future lightcone of the first emission event – and also correlations between the field and the earlier state of the source – but no correlations associated with the second pulse. In fact, as far as the second emissions event is concerned, we can treat the fields as being uncorrelated.

Now, in previous chapters we have already seen the close connection between causal representations and an assumption of initial randomness. An acyclic, deterministic causal model satisfies the Markov condition exactly if the exogenous variables are probabilistically uncorrelated. Thus, if it is possible to represent the relationships among charged particles or between particles and fields in terms of asymmetric causal structures, then the randomness assumption to which Earman appeals ensures that we can apply standard common-cause reasoning to infer the earlier existence of a radiating source from correlations among distant field regions. Thus, both possible explanations of the absence of miraculous coherently converging radiation identified by Earman underwrite the possibility of representing the relation between sources and fields causally. Moreover, just as one might want to explain the successful application of causal structures by appealing to an initial randomness assumption, one can explain the randomness assumption by appealing to the absence of a common cause of the uncorrelated events or explain the absence of truly source-free incoming radiation by pointing to the absence of any sources as cause of such radiation. This equivalence allows Judea Pearl to characterize the initial randomness assumption as itself a causal assumption (see Pearl 2000). Finally, as I argued in the last chapter, it need not be the case that there is uniquely correct answer to the question as to whether the initial randomness assumption or a causally asymmetric relationship between sources and fields is more fundamental. The conclusion of the present discussion I want to stress is merely that once we introduce an asymmetric probabilistic assumption, as Earman does, we also allow for legitimate causal representations of the relation between sources and fields.

Earman's sharp criticism of appeals to causal notions in connection with the radiation asymmetry might be partly motivated by qualms about a rich notion of causation as metaphysical production. And, indeed, I do not think we can draw weighty metaphysical conclusions from the preceding discussion. Yet if we restrict our attention to a metaphysically thinner, functional notion of causation, then it is utterly mysterious why a probabilistic independence assumption ought to be scientifically legitimate but an appeal to acyclic deterministic causal structures is not.

I want to end my discussion by briefly returning to the disagreement between Ritz and Einstein. Here we can distinguish Ritz's disagreement with the Einstein of (1909a) from that with the Einstein of (1909b). Against the kind of view expressed by the later Einstein, Ritz argued that the asymmetry of radiation could not be captured within the context of a field theory. But Earman's probabilistic independence assumption allows us to express an asymmetry even within a field-theoretic framework for finite systems: There is no coordinated behavior of the incoming fields correlated with sources in the future of the initial-value surface. By contrast, there will in general be outgoing fields correlated with the sources in the past of a final-value surface. And this asymmetry can be explained *both* with Ritz by positing elementary actions at a distance *and* with Einstein by appealing to an asymmetry of the elementary field-theoretic processes of radiation.

As far as Ritz's debate with the earlier Einstein is concerned, a rapprochement seems possible. Ritz ultimately appeals to what amounts to a time-asymmetric causal relation between different particles. Einstein, by contrast, argues that the asymmetry of radiation can be given an explanation in terms of a probabilistic constraint – presumably the very condition invoked by Earman. But once we represent the correlations between charged particles in terms of causal structures, it turns out that the probabilistic independence assumption and the fact that we can use information about correlations as input in paradigmatically causal reasoning are two sides of the very same coin. The "scientifically respectable" claim, to say it with Earman, is that the relations between charged particles (and fields) can be represented in terms of causal structures that satisfy the causal Markov condition. For this claim there is ample scientific evidence, and the claim is legitimated by the important role causal reasoning plays in physics.

CHAPTER 8

"Entropy accounts" of causation

I. Introduction

In previous chapters we have repeatedly encountered the fact that there is a close connection between time-asymmetric causal relations and an assumption of initial randomness. This connection is, for example, embodied in the causal Markov theorem, which says that any deterministic acyclic causal model with independent exogenous variables satisfies the causal Markov condition and hence allows for common cause inferences. Conversely, as I suggested, the principle of the common cause can be used to motivate an initial independence assumption: initial states are distributed randomly precisely when they do not have a common cause in their past. Just such an independence or randomness assumption also plays an important role in certain accounts of the thermodynamic asymmetry. In the present chapter I want to contrast my own account with accounts that attempt to ground the causal asymmetry in thermodynamic considerations. The two that I want to examine here are Barry Loewer and David Albert's neo-Boltzmannian account and Huw Price's perspectival account of the causal asymmetry. The overarching difference between these two accounts and my own is that they are reductive. Both take the world to be fundamentally non-causal and the fact that time-asymmetric causal notions are nevertheless useful for beings like us is seen as following from thermodynamic features of the world. My account, by contrast, is non-reductive. Both the causal asymmetry and an initial randomness assumption, in my account, are two aspects of a fundamental temporal asymmetry in the world that is reflected in our explanatory practices and in the representational resources we use.

I will proceed as follows. In the next section I will briefly describe the role of probabilistic assumptions in accounts of the thermodynamic asymmetry. In Sections 3 through 5 I will criticize various aspects of Albert and Loewer's view. The overall conceptual structure of the view is this:

the core assumptions of a statistical account of the thermodynamic asymmetry, including a probability postulate, are meant to underwrite, via an asymmetry of records, a counterfactual asymmetry, which in turn provides the foundation of our causal judgments. My main criticism will be that the account does not result in a suitably sharp temporal asymmetry to ground the causal asymmetry. In Section 6 I will show that the temporal asymmetry of records can be directly derived from the probability postulate alone. In Section 7, finally, I critically examine Price's perspectival account of causation. I end with a brief conclusion.

2. The micro statistical account

According to the Boltzmannian account defended in (Albert 2000), the thermodynamic asymmetry that the entropy of a closed macroscopic system never decreases can be explained by appealing to a time-symmetric micro dynamics and an asymmetric constraint on initial conditions. If we assume an equiprobability distribution of micro states compatible with a given macro state of non-maximal entropy, then it can be made plausible that (intuitively) "most" micro states will evolve into states corresponding to macro states of higher entropy. However, if the micro dynamics governing the system is time-symmetric, then the same kind of considerations also appear to show that, with overwhelming probability, the system evolved *from* a state of higher entropy. This undesirable retrodiction, which is at the core of the *reversibility objection*, can be blocked if we conditionalize the distribution of micro states not on the present macro state but on a low-entropy initial state of the system. Since the reversibility objection can be raised for any time in the past as well, Albert and others argue that we are ultimately led to postulate an extremely low-entropy state at or near the beginning of the universe. Thus, as Richard Feynman concludes in a very readable and easily accessible presentation of this view, it is "necessary to add to the physical laws the hypothesis that in the past the universe was more ordered, in the technical sense, than it is today" (Feynman 2001, 116). This temporally "lopsided" hypothesis, Feynman says, is needed to understand and make sense of irreversibility. Albert and Loewer call Feynman's hypothesis of a low entropy initial state of the universe "the past hypothesis" (*PH*).

Positing an equiprobability distribution at some initial time, however, seems to lead to the following problem. If we postulate a uniform probability distribution over the initial state of a system, then the distribution will not be uniform over micro states compatible with the actual macro state at later times. If later macro states have higher entropy, they will correspond to regions of phase space that are vastly larger than the region corresponding to the low-entropy initial state. But, according to Liouville's theorem, regions of phase space evolve into regions of equal size. Thus, positing an equiprobability distribution at the initial time precludes that the distribution is uniform over macro states at later times and, hence, might seem to preclude us from bringing to bear the very considerations that seemed to ensure that entropy is overwhelmingly likely to increase in the first place.

This problem can be solved, if we assume that the phase space region corresponding to the initial macro state dynamically evolves into a highly fibrillated region such that the micro states that have evolved from the initial macro state eventually are homogeneously distributed over all measurable subregions of the system's available phase space. A formal condition that ensures that this assumption is the condition that a system be *mixing* (see, e.g., Uffink 2006). A dynamical system is a tuple $<\Gamma, A, \mu, T>$, where Γ is the system's phase space, A is the set of measurable subsets of Γ, μ is a probability measure, and T is a one-parameter group of transformations T_t that represents the evolution operators. A dynamical system is mixing exactly if, for all $A, B \in A$,

$$\lim_{t \to \infty} \mu(T_t A \cap B) = \mu(A)\mu(B).$$

For such a system, the micro state at t will with overwhelming probability be "typical" of the micro states compatible with the macro state at t, in the sense required for the Boltzmannian account.

Thus, the assumptions of the statistical mechanical account (SM) from which the thermodynamic asymmetry is derived are the following:

(i)   time-symmetric, deterministic dynamical micro laws.

(ii) the *past hypothesis PH*, which characterizes the initial macro state of the universe as a low-entropy condition satisfying certain further symmetry conditions.

(iii) a *probability postulate PROB*, which postulates a uniform probability distribution over the physically possible initial micro states of the universe, compatible with the past hypothesis *PH*.

(iv) an assumption of mixing or dynamic instability of possible micro evolutions.

This globalist account of the entropy-asymmetry, which aims to derive the thermodynamic asymmetry from assumptions about the early state of the universe, has been challenged (e.g., by Winsberg 2004; Earman 2006), but

I do not want to discuss these criticisms here. In what follows I will assume that the account can successfully explain the thermodynamic asymmetry and ask whether it can be extended to explain the causal asymmetry as well.

Albert and Loewer maintain that the statistical account provides us not only with the correct explanation of the second law of thermodynamics, but with a fundamental theory of the world. The account's core assumptions, Loewer says, provide us with a "probability map of the universe since they entail a probability distribution over the micro histories of the universe compatible with [the initial low entropy macro state] M(o)" (Loewer 2012a, 124). Adopting a term from a movie by the Coen brothers, Albert and Loewer call this statistical-mechanical theory of everything "the Mentaculus": "The Mentaculus is imperialistic since it specifies a probability distribution over all physically possible histories and hence a conditional probability over all pairs of (reasonable) macro propositions" (Loewer 2012b, 18).

What is the status of the "lopsided" hypotheses *PH* and *PROB* in the account? Albert and Loewer argue that these hypotheses, which need to be added to the laws, are themselves nomic constraints, offering the following three reasons for this view.[1] First, it is a desideratum that thermodynamic principles such as the "second law" have the status of laws (even if only probabilistic laws). Since the second law is, according to the *SM* account, derived from *PROB* and *PH*, "it is absolutely essential," as Loewer says, "that *PROB* be understood as a law if it is to ground the increase of entropy as lawful" (Loewer 2008).

Second, if *PH* and *PROB* are treated as laws, they can provide a crucial missing piece in a broadly Lewisian account of counterfactuals and causation. Lewis famously attempted to derive the temporal asymmetry of counterfactuals and causation from a thesis of an asymmetry overdetermination, according to which multiple localized facts about the present are nomologically sufficient for the occurrence of events in the past, but that the future is not similarly overdetermined by the present (Lewis 1979): there are many events that have multiple determinants at a given time in their future, that is, many different sets of minimally sufficient conditions for the event, but events do not similarly have multiple determinants at a given time in their past. But Lewis's overdetermination thesis is provably false. No separate local facts at one time are individually nomologically sufficient for events at other times – rather, the laws require the state on

---

[1] Given the strong similarities in the views they express in print, I shall here for ease of exposition assume that the views defended in papers authored by Loewer or Albert alone also express views held by them jointly.

a complete initial- or final-value surface as input to determine the state of a system at other times; and in the case of time-symmetric laws, there is no *asymmetry* of determination.[2] Thus, any attempt to rescue a broadly Lewisian account of a counterfactual asymmetry needs to supplement the account with explicitly time-asymmetric assumptions. Loewer argues that adding *PH* and *PROB* as *nomic* constraints does the job and allows us to derive a counterfactual asymmetry in a non-question-begging way.

On Loewer's account, we evaluate counterfactuals by calculating the probability of the consequent, *conditional on the laws of the actual world*, the counterfactual antecedent event *c* at some time *t*, and the actual macro state at *t* outside of the region where *c* occurs. The asymmetry of counterfactuals is then a consequence of the fact that the laws include the time-asymmetric constraints *PH* and *PROB*. If instead of treating *PROB* and *PH* as laws, we merely imposed them as de facto asymmetric constraints on the past evolution of counterfactual worlds (in addition to the dynamical laws), we would violate Lewis's desideratum of *deriving* the counterfactual asymmetry rather than merely putting it in by hand. The difference is that, if *PROB* and *PH* are nomic constraints, then the proper procedure for evaluating counterfactuals can be characterized in an apparently non-question-begging way as that of conditionalizing on all the laws of the actual world, which just happen to include lopsided time-asymmetric constraints.

Thus, Albert and Loewer argue that the Mentaculus implies a temporal asymmetry for a certain kind of counterfactual, which they in turn take to underwrite our causal judgments. By contrast, on the view I am defending in this book, a probabilistic independence assumption and the causal asymmetry are interderivable, and both serve to underwrite an asymmetry of causal counterfactuals. As we will see, the attempt of grounding the causal asymmetry in a counterfactual asymmetry is one of the main sources of difficulties for Albert and Loewer's account.

The third reason for why *PH* and *PROB* have the status of laws is that, according to Loewer, they are axioms of the Lewisian Best System. According to the Mill-Ramsey-Lewis (MRL) account of laws, we can represent the totality of our scientific knowledge of the world as having the structure of a deductive system consisting of a set of axioms and of all the axioms' deductive consequences. Various deductive systems may differ in their

[2] See Frisch (2005a, ch. 7) for an argument to that conclusion. An overdetermination thesis similar to Lewis's is defended by Daniel Hausman (1998). Hausman's argument is criticized in Frisch (2005, 185–187) and in Schurz (2001), where Gerhard Schurz shows that the purely probabilistic relations governing intervention variables do not exhibit a temporal asymmetry.

deductive strength and in their simplicity. A system's deductive strength consists of how many truths it contains, whereas a system's simplicity is a measure both of how many independent axioms it contains and of how syntactically simple these axioms are. Deductive strength and simplicity are competing criteria. According to the MRL account, the laws are those generalizations that are axioms of the deductive system that strikes the best balance between simplicity and strength.[3]

In Lewis's version of the view, the laws must be formulated in terms of "natural predicates" that pick out fundamental properties. Albert and Loewer's version of the view allows for less fundamental predicates – in particular for thermodynamic predicates – since adding these predicates greatly simplifies the deductive system. The resulting account further amplifies the pragmatic element in the MRL account that arguably is already present in Lewis versions. The pragmatism is brought out particularly vividly in an imaginary tale of how the Best System is revealed to us during an audience with God. Here is how Albert describes the scenario. Imagine that you have an audience with God, who provides you with as much information about the particular facts of the worlds as you could possible want to have. One way to provide this information is to recite long lists of particular facts concerning which properties are instantiated at which spatiotemporal locations. Yet as God begins to recite the fact,

> it begins to look as if all this is likely to drag on for a while. And you explain to God that you're actually a bit pressed for time, that this is not all you have to do today, that you are not going to be in a position to hear out the whole story. And you ask if maybe there's something meaty and pithy and helpful and informative and short that He might be able to tell you about the world which (you understand) would not amount to everything, or nearly everything, but would nonetheless still somehow amount to a lot. Something that will serve you well, or reasonably well, or as well as possible, in making your way about in the world. (Albert unpublished)

The meaty and pithy information with which God provides you, Albert and Loewer claim, consist of the micro dynamical laws together with *PH* and *PROB*. The Mentaculus provides the best account, because it combines simplicity and strength in ways that are most useful and best *for us*.

Thus, the ultimate yardstick for simplicity and informativeness is how practically useful a system is for us – *how well it allows us to make our way about in the world*. For beings like us, the deductive system that includes the

---

[3] For a critical examination of Loewer and Albert's version of a best-system account of laws, see Frisch (2011b and 2014).

PH is clearly simpler – even though from the perspective of the language of the fundamental micro theory, stating the *PH* would be a "gruesome mess." By comparison, a system consisting of the laws and the exact initial conditions would fail dramatically as far as its usefulness for us in making our way about in the world is concerned, since it would be much too complicated to be of any practical use.

Indeed, specifying the exact initial conditions of the universe, Albert says, would violate the stipulation of providing a simple summary: "I can't tell you exactly what that [initial condition of the universe] was," God says in Albert's story, "It's too complicated. It would take too long. It would violate your stipulations." That is, God does not offer a comparative assessment telling us that the loss of simplicity of adding the exact initial conditions of the universe would not be made up by a gain in informativeness. Instead, She tells us that the exact conditions would violate a minimal condition of simplicity. Thus, practical usefulness provides not only a criterion of *relative* goodness for a system but also a *necessary* condition for being minimally acceptable: a system that includes axioms that are too complex and violate a minimal standard of simplicity is practically useless and hence could not be the Best System, no matter how informative it might be in principle or how much more informative it might be than any of its competitors.

3. From the thermodynamic asymmetry to a branching tree structure?

Loewer argues that the Mentaculus underwrites the counterfactual asymmetry since it entails that possible macro histories of the world exhibit a certain tree structure: even though the micro history of the world is assumed to be deterministic, the evolution of macro histories is future-indeterministic in that more than one future macro history will in general be compatible with the macro state of the world at a time and have a probability non-negligibly different from zero. This contrasts with the probabilities assigned to different past evolutions:

> From a typical macro state in the middle of the actual macro history there will be branching in both temporal directions but there will be much more branching where the branches have substantial probability in the direction away from the time of the *PH* than back towards it. The overall structure is due to the fact that the macro state at *t* (in the middle) must end up in the direction of the boundary condition at which *PH* obtains (the direction we call 'the past') satisfying *PH*. (Loewer 2007, 302)

And:

> [*PH* and *PROB*] determine an objective probability distribution over all
> nomologically possible micro histories (and *a fortiori* over all macro histories
> and all macro propositions). Even though the underlying micro dynamics
> is deterministic macro histories form a tree structure branching towards the
> future (away from the time at which *PH* holds). (Loewer 2007, 307)

That is, the objective probability distribution determined by *PH* and *PROB*
forms a branching tree structure – a tree structure that is due to the fact
that *PH* provides a constraint on possible evolutions.

We can express this structure somewhat more formally by introducing
the notion of quasi-determinism:

> (QD) A system is *quasi-deterministic* at t relative to some time t' and some
> set of mutually exclusive macro states M, exactly if there is a state M_i in M
> such that $P(M_i(t')/S(t))$ is close to 1, where S is the state of the system at t.

The probabilities here (and throughout this chapter) are the ones induced
by the statistical mechanical probability distribution and conditionalization
on the dynamical micro laws and the past hypothesis *PH* is left implicit. The
claim that the universe exhibits an asymmetric tree structure is equivalent
to the conjunction of the following two claims:

(1)   The world is not quasi-future-deterministic; or more precisely: for all
      times t there is a Δt, such that for all times $t' > t + \Delta t$ and all M,
      the world is not quasi-deterministic at t relative to t'.

(2)   The world is quasi-past-deterministic at all times t with respect to all
      times t', $t' < t$, and all M.

As I show in Frisch (2005c), it is indeed possible to derive from a formal-
ization of the tree structure similar to QD an asymmetry for the type of
counterfactual that is at the core of Loewer's account. Yet the claim that the
Mentaculus implies a tree structure of possible macro evolutions is highly
problematic, as I want to argue now.

Since macro states closer to equilibrium occupy vastly larger regions of
phase space than states further away from equilibrium, it follows from Liouville's theorem that there will be many possible different non-equilibrium
states far from equilibrium that evolve into the same state closer to equilibrium in the future. This suggests that there may be many more changes
to the micro state of a system close to equilibrium *associated with different
macro pasts further away from equilibrium* than there are changes to the
micro state of a system far from equilibrium *associated with different macro
futures closer to equilibrium*. Merely comparing the phase-space volumes

associated with macro states at different times suggests that possible macro evolutions may exhibit an upside-down tree structure.[4]

We can distinguish two worries here: first, focusing on the future "end" of the tree structure, is it indeed a consequence of the Mentaculus that there will be no significant reconvergence of branches, and that there are no times with respect to which thermodynamic systems are quasi-deterministic? And, second, focusing on the past "end," is it a consequence of the account that the past is quasi-deterministic at all times with respect to the initial time t_{PH} at which PH holds?

That future evolutions are not quasi-deterministic might seem to follow from the assumption of mixing. If a system is mixing, the conditional probability $P(M(t)/M_o)$ of a macro state $M(t)$ given the initial state M_o depends only on the phase-space volume associated with $M(t)$ and is independent of M_o. Yet the mixing assumption alone does not imply the failure of quasi-determinism for all future times. If there is a single equilibrium macro state M_e that takes up the overwhelming majority of the phase space region available to a system, then $P(M_e)$ can be close to 1 and the system is quasi-future-deterministic with respect to all times after which the system reaches equilibrium.

This point holds for thermodynamic systems of all sizes – to the extent that the Boltzmannian account applies to these systems – ranging from small macroscopic quasi-isolated systems to the universe as a whole. Consider, for example, *the* paradigmatic thermodynamic system – a body of gas: assume that the gas is initially confined to the right half of a container and, after a partition is removed, spreads out until it is distributed evenly throughout the container. Since most of the phase space accessible to the gas is associated with its equilibrium state, the Mentaculus allows us to predict that the gas will be overwhelmingly likely to end up in that state – the gas evolves quasi-deterministically with respect to the final equilibrium state. At the other extreme, current cosmology suggests that the universe as a whole, too, may be quasi-deterministic with respect to its future equilibrium state, in which ionized stable particles, that is, protons, neutrons, and electrons,

---

[4] Since Loewer represents possible macro evolutions in his diagram of a possible tree structure by cylinders of constant diameter, the diameter cannot be taken to represent phase space volumes. If we wanted to include representations of the phase space volumes associated with macro states in the diagram, possible macro evolutions would have to represented by cones of dramatically increasing widths toward the future. As a rhetorical device, Loewer's diagram lends far more plausibility to the thesis of macro branching toward the future, than a picture of cones that branch at the same time as they dramatically increase in width. (When you try to draw this, you'll quickly run out of space into which branching could occur.)

are distributed evenly throughout the cosmos at a density approaching zero (see, e.g., Baez 2011).

Although Loewer is obviously right in suggesting that there are many systems that are open to the future – there clearly is widespread macro branching toward the future – thermodynamic considerations imply that there also is widespread reconvergence of possible macro histories. Thus, at the cosmological level, even though the initial state of the universe might not determine the large-scale distribution of matter before elementary particles begin to "boil off" in the final evolution toward equilibrium, different cosmological macro histories will converge toward the final equilibrium state. Similarly, there is convergence at the level of "human-sized" macro systems: no matter which part of the container a body of gas occupies initially, after the partition is removed the gas will spread until it is uniformly distributed throughout the container.

As a simple case exhibiting both branching and reconvergence, consider the example Albert uses to motivate postulating a past hypothesis and the existence of macro branching (Albert 2000, 82ff.). Albert imagines a system consisting of ice cubes that drop into glasses of water after sliding down a device similar to a Galton board. The same low-entropy initial state, with the ice cubes collected at the top, will indeterministically evolve into different macro states given by different distributions of ice cubes in the glasses at the bottom of the board. Yet if we imagine that the ice cubes have several macroscopically distinct shapes of the same volumes, then there will be macroscopically distinct distributions of ice cubes in the glasses that will eventually evolve into the same macro states once the ice is fully melted. And if we further assume that at the end of our experiment all glasses with water are emptied into a single bucket, all possible macro histories that diverge after the ice is released a the top of the Galton board will reconverge – no matter what the shapes or volumes of the ice cubes are and no matter what path they take down the board. That is, the final state of the system when the all the water is collected in the bucket is not quasi-deterministic with respect to *past* times when the ice cubes were distributed among the different glasses, even if we impose as additional constraint that all macro histories are constrained to have originated in the state where the ice cubes were collected in a container at the top of the Galton board.

Now, there are discussions in the literature on counterfactuals that suggest that there is a crucial consideration that has been missing from our examination so far – the role of records or traces of the past. These discussions often invoke Kit Fine's famous example of Nixon's pushing the button that leads to a nuclear holocaust. It is often suggested that the many traces Nixon's action (or inaction) leave in the world play an important role in making convergence of "button-pushing worlds" with "non-button-pushing worlds" difficult. In the case of the ice cubes sliding down the Galton board, drops of water on the board or my memories of observing a particular ice cube slide down a certain path might constitute such traces.

But it is easy to exaggerate how frequent and persistent macro traces are. In fact, it is precisely the thermodynamic behavior of systems that often either prevents the formation of macro traces or leads to the disappearance of such traces. Whatever else the connection between *PH* and the existence of records is, one central role played by the thermodynamic arrow is that of the great destroyer of macro records and macro traces. Thus, any drops of water left on the Galton board, which constitute traces of an ice cube's trajectory, will eventually evaporate; and since which path a particular ice cube took does not have the same momentous consequences for Earth's fate as Nixon's decision whether to push the button, I will soon forget any details of what I observed (and may well forget altogether that I ever conducted the experiment). Nor will there be any other macroscopic "traces" of the experiment. Although light waves will be reflected differently by the ice cubes depending on their path because of the multiple scatterings of photons off of laboratory walls and air molecules, these differences will leave no macroscopic traces by the next day. We might even imagine that there are different lamps that light up depending on what path an ice cube slides down. By the next day – and in fact much sooner – there will be no macroscopic traces of a particular lamp's having been lit when ice cube 17 slid down the board. Because of the thermodynamic behavior of the walls of the laboratory and of the atmosphere, the macro state of the world tomorrow will be completely independent of what the outcome of my experiment is today.

Similar examples can be multiplied indefinitely. Although there indubitably are many systems that for some finite time do *not* evolve quasi-deterministically, there are also many cases like the ones I just described – cases for which differences even in the current macro state will eventually "wash out," for which the system's macro history throughout some period *T* will leave no macroscopic traces in the future, and for which different macro states will evolve quasi-deterministically into one and the same future macro state. Even if possible macro evolutions of *some* systems at *some* times exhibit the kind of branching that Loewer postulates, this behavior is not ubiquitous enough to be able to underwrite a counterfactual asymmetry general enough to be able to ground the causal asymmetry.

I have argued that the assumption of mixing is not enough to ensure that a system is future quasi-indeterministic with respect to its equilibrium state and that it is a consequence of the thermodynamic behavior of systems that there will be reconvergence of possible macro histories even for systems that do not evolve quasi-deterministically during some time interval T. Can mixing at least ensure that the evolution *toward* equilibrium is not deterministic? It is far from clear that the answer is "yes." First, all we can conclude from the assumption that a system is mixing is that after a sufficiently long time the probability of finding a system in a given macro state is proportional to the phase-space volume associated with that state. That is, we can conclude from the fact that a system is mixing that it will *end up* in an equilibrium state, but we cannot draw any inferences at all about *how it will get there*.

Second, as Earman (2006) has argued, if we were able to show that all thermodynamic macroscopic systems had different possible macro futures that receive substantial probability, we might be showing too much, as it were, and our theory would be empirically inadequate. Although there clearly are systems that are dynamically unstable on the macro level, there also are many systems that do not exhibit any macroscopic instability and are quasi-future-deterministic. Indeed, many paradigm cases of causal or time-asymmetric counterfactual judgments concern such quasi-deterministic macro systems. Not only might we want to endorse the claim that had the proverbial butterfly not flapped its wings, there would not have been a storm – an example of a causal counterfactual concerning a dynamically unstable system – but we might also want to say that had I not stepped on the brake, my car would not have come to a halt at the red light – an example of a causal counterfactual concerning, one hopes, a quasi-deterministic system. One might worry, then, how we can recover the apparently deterministic macro evolutions of many systems from the assumption of dynamic instability on the micro level.

The picture that has emerged is not one of an asymmetrically branching tree structure, but rather that of a web of possible macro histories that branch and reconverge. Whether at its future end the web of possible macro histories for the universe converges into a single strand is a question for cosmology to decide. But the sub-web characterizing the history of Earth and many of the even lower-dimensional "sub-webs" characterizing human-scale subsystems on Earth involve a large amount of convergence of strands, as well as branchings. Moreover, there are many small macro systems that over (humanly) significant stretches of time evolve quasi-deterministically and do not exhibit any branching at all.

So far I have focused on Loewer's claim that there is branching toward the future without widespread reconvergence. I now want to turn to his claim that it follows from the SM account that the macro evolution of the universe is quasi-past-deterministic with respect to an initial time t_{PH}. Earlier I expressed the past hypothesis as the constraint that the initial macro state of the universe was *a* low-entropy state satisfying certain further symmetry conditions. But if this indeed is what the past hypothesis says, an additional problem arises for Loewer's claim that the past is closed: It does not seem to follow from the constraint that micro histories originated in *a* very low-entropy state that the macro past is *the* unique actual low-entropy past. That is, counterfactual micro histories may have originated in low-entropy macro states *distinct* from the actual low-entropy past. Consider once more a system consisting of a gas in a box and assume that the gas could have started out in one of two possible low-entropy initial states, confined either to the right or the left half of the container. Let us assume that in the actual world the gas started out in the left half of the container and then spread out until it reached equilibrium, filling the entire container. Then, according to the reversibility objection, most changes to the final micro state will be associated with a high-entropy past, since most micro states compatible with the final equilibrium state will have evolved from equilibrium initial states. What if we assume a "past hypothesis" and constrain changes to the final micro state to those that evolved from *a* low entropy initial state? The phase space regions corresponding to the two initial states – the gas confined to the right or to the left half of the box – will evolve into highly fibrillated regions. If we assume that the system is mixing, each coarse-grained "box" of phase space will have the same proportion of points that have evolved from the two initial regions. That is, intuitively, while the overwhelming majority of points in each box of phase space lie on trajectories that have evolved from high-entropy pasts, the same number of points in each box lies on trajectories that originated in the two low-entropy states. Given the final macro state, the system is as likely to have evolved from the non-actual low-entropy past where the gas would have been confined to the right half of the container as from the actual past, and adding a low-entropy constraint in the past does nothing to privilege the actual low-entropy past.

If *PH* merely restricts macro histories to have originated in some (suitably symmetric) low-entropy state, then Loewer's conclusion that the universe is quasi-deterministic with respect to t_{PH} seems unwarranted. But Loewer himself characterizes *PH* differently: He says that *PH* is "a statement *specifying the macro state* of the universe at one boundary" (Loewer 2007, 300, my italics). That is, according to Loewer's reading, the past hypothesis restricts possible micro histories to have originated in *the actual* low-entropy past state, and this restriction trivially ensures that all possible macro histories originated in one and the same macro state. But can we assume the *actual* initial macro state as constraint, without begging the question, in an account that is meant to derive a temporal asymmetry of counterfactuals?

Loewer's explicit aim is to provide a broadly Lewisian account of a counterfactual asymmetry, and he contrasts his and Lewis's strategy, on the one hand, with Jonathan Bennett's, on the other. Bennett does not offer an explanation of the asymmetry but simply assumes that counterfactuals are evaluated by keeping the past fixed. Loewer says:

> I think that Bennett's account does a pretty good job of characterizing a conditional that matches core uses of the counterfactuals that interest us . . . However, Bennett's procedure for evaluating counterfactuals *assumes* the distinction between past and future (since forks are to the future) and so it does not provide a scientific explanation of time's arrows. (Loewer 2007, 309–10)

Thus, in order to provide a scientific explanation of the asymmetry, we cannot merely assume the asymmetry by holding the past fixed and allowing only the future to vary, but must derive this asymmetry from the global distribution of matters of fact in the actual world in conjunction with the laws.

One might worry, then, that by the very fact that Loewer assumes *PH* as a time-asymmetric constraint he, like Bennett, is putting in the asymmetry by hand. Both Bennett and Loewer, it seems, stipulate a time-asymmetric constraint on how past states of the world may vary, and from this derive that counterfactuals are time-asymmetric. To be sure, Bennett's constraint goes beyond Loewer's – he stipulates that we hold fixed the *entire* macro history in one temporal direction, whereas Loewer only fixes the macro state at the past temporal *end* – but Loewer's constraint may strike one as similarly question-begging, if our goal is to provide a scientific explanation of a temporal asymmetry of counterfactuals. The only "scientific contribution" to Loewer's account might be that the dynamical laws need to ensure that counterfactual past micro evolutions converge quickly enough with the actual macro past.

Loewer's reply to this worry is that the initial macro state of the universe plays a special role in our overall scientific conception of the world. Loewer himself, as we have seen, tries to capture this role by proposing a broadly Lewisian account of laws and suggesting that the actual initial macro state is part of the Lewisian Best System. Yet he apparently also believes that the special scientific status of the initial macro state can be motivated independently of Lewis's account of laws. What, then, is the special role played by the *PH*, and does that role provide us with good reasons for assuming the *actual* initial macro state (rather than just *a* low-entropy state) as constraint on possible macro histories?

First, in the Boltzmannian account, *PH* plays a central role in deriving the thermodynamic asymmetry. Thus, Loewer supports affording *PH* a special role by saying that it "underwrite[s] many of the asymmetric generalizations of the special sciences especially those in thermodynamics and these generalizations are considered to be laws" (Loewer 2007, 304). But in order to derive the thermodynamic "laws," it is sufficient to assume that the universe began its life in *a* low-entropy state (in addition to *PROB*). Thus, the foundations of thermodynamics do not provide us with a reason to accept Loewer's version of *PH* as constraint instead of the one I proposed.

Second, both Albert and Loewer point to the explanatory role the actual macro state of the early universe plays in current cosmology. Thus, they maintain that any macro state that results from a small hypothetical alteration to the actual present macro state is constrained to have evolved from the actual initial macro state that cosmology will eventually present to us. But the plausibility of this claim relies on an equivocation on the notion of macro state. A macro state is associated with a coarse-graining over the phase space of a system – and in different contexts, different coarse-grainings are appropriate. In the case of the kind of counterfactuals associated with paradigmatically causal claims, the right level of description is one referring to medium-sized, "human-scale" objects, whose states are characterized in units such as 1 m or 1 kg. In the context of astronomy or cosmology, we are interested in the distribution of stars and galaxies, and demanding that macro states be specified to a precision of the location of medium-sized objects would be absurd. Appropriate units in the latter context are, for example, the astronomical unit $1\,\text{AU} = 1.5 \times 10^{11}$ m or the solar mass $1\,\text{M} = 1.9 \times 10^{30}$ kg. Thus, even if we grant that a specification of the actual initial macro state of the universe provides a scientifically legitimate and non-question-begging constraint on possible macro evolutions, the constraint can only be a specification of the *cosmological*, coarse-grained macro state. Any specification of the initial macro state more fine-grained than that does not play a scientifically explanatory role.[5] But just as there

---

[5] Within Loewer's preferred Lewisian account of laws, this point can be made as follows: Including a description of the universe's fine-grained, human-scale initial macro state in our deductive system will vastly complicate the system without providing us much (if any) gain in informativeness.

are many *micro* states compatible with a given fine-grained macro state, there are many fine-grained macro states (specifying, for example, the exact distribution of small rocks on a planet's surface) compatible with a more coarse-grained macro state.

One might think that the specific nature of the macro state of the early universe provides a reply to this worry. According to current cosmology, matter was distributed smoothly shortly after the Big Bang. (A smooth matter distribution, it is often argued, represents a state of extremely low gravitational entropy, and hence, as matter clumped to form stars and galaxies, the gravitational entropy of the universe increased.) Thus, one might think that there is just a single initial macro state tout court – that is, even just a single fine-grained macro state – that satisfies the conditions revealed to us by cosmology. Whereas there can be many different macro states that exhibit the same amount of gravitational clumping, there seems to be only a single macro state characterized by a completely smooth matter distribution – a state that is smooth at all levels of coarse-graining.

But this reply fails for two reasons. First, its premise is false. The macro state of the early universe was not completely smooth, even on a cosmo-logical level and – fortunately for contemporary cosmology – exhibited density fluctuations large enough to function as seeds for the formation of stars and galaxies.[6] Second, the inference from a distribution that is smooth at one level of coarse-graining to one that is smooth at all levels is not sound. It is part of the Boltzmannian account that the micro state of the early universe was one that is "typical" given the known macroscopic constraints. This means that, if the association between a smooth mat-ter distribution and low gravitational entropy is correct, the Mentaculus implies that the early universe is overwhelmingly probable to have exhib-ited as much gravitational clumping as is compatible with our cosmological evidence.

Thus, neither statistical physics nor cosmology provides us with scien-tific reasons to take the actual *fine-grained* or human-scale macro state as constraint on possible fine-grained macro evolutions. The Boltzmannian account requires as premise only that the universe began its life in *a* state of extremely low entropy, and cosmology restricts that state to one that is characterized by an approximately smooth matter distribution, but with density fluctuations large enough to be compatible with many different fine-grained macro states.

---

[6] The density fluctuations are of the order of 1 part in 100,000. By comparison, differences in mass distribution of interest to us are of the order of 10^{-30} times the mass of the sun.

4. Causal handles

In the last section I have argued that it is a consequence of the thermodynamic arrow that there will be convergence among different possible macro evolutions and that there will be many cases where small differences in the macro state of a system at one time leave no macro traces in the system's future. In this section and the next I will show that this result leads to problems both for Albert's of causal handles and for Loewer's account of decision-counterfactuals.

Albert argues that it is a consequence of the Boltzmannian account that the present contains multiple *causal handles* on the future but (almost) no causal handles on the past. If we constrain the *remote* past of any physical system, he maintains, then only very few and special alterations to the present are associated with a different *recent* past, while many such alterations may lead to different futures. To illustrate this point, Albert asks us to consider a collection of idealized billiard balls on a frictionless plane such that ball 5 is currently stationary with the additional constraint that ball 5 was moving 10 seconds ago. Given this additional constraint, the fact that ball 5 has been involved in a collision in the past 10 seconds is nomically *determined* by facts about the present state of ball 5 *alone*. That is, alterations to the present state of the balls *not* involving changes in the state of ball 5 cannot change the fact that ball 5 was involved in a collision during the last 10 seconds. Yet there are many changes to the state of the balls not involving ball 5 that will result in a different future evolution of ball 5. From this Albert concludes that there are a far wider variety of "what we might call *causal handles* on the future of the ball in question here, under these circumstances, than there are on its past" (Albert 2000, 128). In this example the constraint that ball 5 was moving is meant to play the role of a "past hypothesis," and the current state of ball 5 functions as a record of the past collision. More generally, then, Albert claims that, if we postulate *PH* as constraint on all possible macro evolutions, then this imposes almost no additional restriction on possible future macro evolutions, while it restricts non-actual present macro states that are the result of small macro changes to the actual present state to have evolved from the *actual* macro past – that is, it follows from imposing *PH* as constraint on all possible macro histories that there are many more causal handles on the future than on the past.

In Albert's example the current state of ball 5 together with the past constraint nomologically determine that ball 5 was involved in a collision. In the general case, however, *PH* in conjunction with certain local facts about the current macro state assigns probabilities strictly less than 1 to the occurrence of past events. Many records or traces of the past do not determine the occurrence of the events of which they are records but only raise the probabilities of their occurrence. Thus, the general definition of a causal handle is as follows: A macro event $C(t)$ is a *causal handle* on an event $E(t')$ exactly if the occurrence of C (significantly) affects the probability of E. That is, $C(t)$ is a causal handle on $E(t')$ exactly if $P(E/C \& M(t)) \neq P(E/M)$. $M(t)$ is the actual macro state at t outside of the region where C occurs and contains any putative records of E at t. In Albert's proposal, we evaluate the results of small hypothetical changes to the present by keeping the present macro state fixed except for the small change and then determine how this counterfactual macro state evolves in accord with the constraint given by the Mentaculus – with one important qualification: Albert assumes that, in addition to any macro records of an event, we also hold fixed any putative *memories* we might have of that event, even though memories might be physically realized by micro states.

Albert's thesis that there are (almost) no causal handles on the past is tantamount to a screening-off condition. $C(t)$ is not a causal handle on some past event E exactly if the rest of the macro state at t screens off E from C – that is, $P(E/C \& M(t)) = P(E/M)$. But for events E that leave at most a small number of traces in the present, this condition can easily fail. Take an event $E(t')$ that has only two distinct macro traces C_1 and C_2 at some later time t. C_1 and C_2, intuitively, are both effects of E. Thus, although E as the common cause of C_1 and C_2 might screen off C_1 from C_2, it will not in general be the case that Albert's condition is satisfied and that one effect screens off the cause from the other effect. Indeed, the presence (or absence) of additional traces of an event – that is, of additional evidence for the event's occurrence – can radically alter the probability of that event. Albert's Galton board can again serve as an example. That a particular ice cube landed in the second glass from the left, say, constitutes a trace of it having slid down to the left of the first pin. (For a board with n rows of pins where the probability at each pin of the ice cube sliding down on one side is equal to $1/2$, this probability is $P(\text{left/second glass to left}) = 1 - 1/n$.) Now let us imagine that the ice cube can dislodge a little ball as it slides down the board and that where the ball ends up also functions as a probabilistic record of the ice cube's path down the board. It is then easy to set up the probabilities in such a way that both the little ball's present position and the ice cube's landing in the second glass come out as causal handles on the path of the ice cube past the first pin. That is, it is easy to set things up such that (keeping the present condition of the ice cube in the glass fixed) there can be many alterations in the present condition of the little ball that would alter the probabilities about whether or not the ice cube slid down to the left of the first pin.[7]

Again, there is nothing unusual about this example. There are many cases where additional evidence affects the probabilities of past events, and hence, according to Albert's account, would count as a causal handle on the past. It seems to me that to the extent that Albert's thesis may appear intuitively plausible, this rests on at least one of the following two assumptions. First, Albert's thesis is true if we demand that a trace (together with the past hypothesis) nomologically *determines* the event's occurrence. Or second, it is true if we demand that each event leave sufficiently many and varied traces that each trace taken individually only marginally affects the probability of the event's occurrence. But as I have argued in the last section, it is a consequence of the thermodynamic arrow that this assumption is often false. There are many mundane (and paradigmatically causally related) events that leave no or only very few traces in the future.[8]

5. Decision counterfactuals

By contrast with Albert's account, Loewer's account of an asymmetry of control focuses primarily not on a purely physical asymmetry but on an asymmetry involving agents. Loewer argues that the Mentaculus account underwrites an asymmetry of *decision counterfactuals*. A decision counterfactual is a probabilistic counterfactual of the form "If at t I were to decide D, then the probability of B would be p," which is true exactly if $P(B/M(t) \ \& \ D(t)) = p$. $M(t)$ is the complete macro state at t and the decision $D(t)$ is an event "smaller than a macro event but with positive probability" (Loewer 2007, 316). A property of decision events that is attractive from Loewer's perspective is that small differences in a decision state can be magnified into large macroscopic differences in the world, and he maintains that this feature of "decision conditionals [is] temporally asymmetric": "Alternative decisions that can be made at time t typically can make a big difference to the probabilities of events after t . . . but make no difference to the probabilities of macro events prior to t" (317).

Trying to capture the idea that different decisions are "open" to an agent at a time, Loewer assumes that decisions are "indeterministic relative to the macro state of the brain and environment prior to, and at the moment

---

[7] And this is meant as an explicit contrast with what Albert says about the billiard ball case on the top of page 127 in his book (Albert 2000).

[8] I critically examine several other aspects of Albert's account of the causal asymmetry in Frisch (2007).

of making the decision" (317). From this assumption, it seems, there is an extremely quick argument for the asymmetry of decision counterfactuals. If the assumption is understood not merely as denying determinism but as asserting that decisions are probabilistically independent of the macro state prior to t, it directly follows that differences in decisions "make no difference to the probabilities of macro events prior to t." But this argument for the asymmetry of decision counterfactuals does not rely on the Mentaculus at all and seems question-begging – the asymmetry of decision counterfactuals is simply built into our account of what a decision is. If we want to avoid begging the question, we need to treat Loewer's decision counterfactuals analogously to Albert's causal handles: in evaluating the truth of a counterfactual, we hold the actual present *macro* state and, in addition, our present memories fixed, posit an alternative decision-event compatible with the state we keep fixed, and then let the conditional probabilities of both future and past macro events be those given by the Mentaculus. For the account to succeed, Loewer's thesis that alternative decisions make no difference to the probabilities of past events would have to come out as a consequence of the Mentaculus.

Yet the Mentaculus fails to imply Loewer's thesis. As a matter of fact, our decisions at t are not completely independent of the macro state of the world prior to t – many of my decisions today reflect facts about my biography and are strongly correlated with past experiences. While there may be decisions that amount to mere random "picking" and hence may be probabilistically independent of my past, many of my decisions exhibit a certain coherence and represent facts both about who I am and about the world.[9] That is, for many of my decisions there are events B in the past such that $P(B/D(t)) \neq P(B/not\text{-}D(t))$. Moreover, acknowledging this dependence does not force us to deny Loewer's assumption that different decisions or choices are "open" to an agent making a decision, since plausibly this assumption can be captured by supposing that an agent's beliefs and desires do not determine her choices (see Holton 2006, 4), and this supposition is compatible with the claim that an agent's choices are probabilistically correlated with events in her past. Finally, even though my history plays a role in shaping the choices I make, I consciously remember only very few events of my past, and only very few of these events have left completely reliable macroscopic traces in the present. Thus, the present macro state in conjunction with my memories does not screen off the past from my decisions – that is, for

---

[9] For a discussion of the distinction between "picking" and "choosing," see Holton (2006).

many of my decisions it will be the case that there are past events B such that $P(B/M(t)\ \&\ D(t)) \neq P(B/M(t))$.[10]

Thus, many of an agent's decisions do make a difference to the probabilities of macro events prior to the time of her making the decision. Now, Loewer argues that even in these cases there still is an important asymmetry between past and future correlations, since, he maintains, we cannot have *control* over events in our past. According to Loewer, the condition of having control is strictly stronger than the condition of probabilistic dependence: "control by decision requires that there be a probabilistic correlation between the event of deciding that p be so and p being so and one's knowing (or believing with reason) that the correlation obtains" (Loewer 2007, 318). Loewer's first condition on control is that there has to be a probabilistic correlation between a decision D and the event B over which the decision maker has control. This condition, I have argued, is satisfied for large sets of pairs of decisions and events in their past.

A second condition is that we must have good reasons to believe that such a correlation obtains. This condition, too, appears to be frequently satisfied, since we are often in a position to discover how our decisions are correlated with our history. Richard Holton argues that one important role for decision or choice in our lives might be that it enables agents to come to know something about themselves *and* about the world that they would not have been in a position to know prior to their decision (Holton 2006). According to Holton, by looking at their choices, agents "can form, rather than just discover, their judgments on that basis" (Holton 2006, 10–11). In Holton's account, an agent who has the right kinds of competences can in certain circumstances learn from her decisions, since her decisions can act like a finely tuned instrument that picks up on cues that are not consciously available to the agent. If some account like this is correct, then there are correlations between our decisions and past events that we can come to believe with reason, and Loewer's second condition is satisfied as well for certain past events.

Thus, it seems possible for us to come to learn of strong correlations between actions or decisions of ours of a certain kind and past events that trigger them. Thus, that we have performed a certain type of action (or

---

[10] As a putative counterexample to his account, Loewer asks whether my decisions now can affect the existence of Atlantis. One implication of my discussion here is that this is not the kind of counterexample about which Loewer should worry. Much more worrisome than the case of Atlantis for Loewer's account are events in the past that the agent facing a decision experienced, but that left (almost) no traces in the present.

made a certain decision) can provide us with good evidence for the occurrence of its past trigger, even if on that particular occasion the occurrence of the trigger has left no traces other than the fact that we made a certain decision. Given my decision, I can have good reasons to believe that its past trigger occurred. Yet, according to Albert and Loewer's counterfactual account, the past trigger comes out as an *effect* of the action rather than as one of its *causes*.

Yet one might think that while there are many cases where one of the two conditions is satisfied individually, the two conditions can never be jointly satisfied when the events in question lie in the past of a decision. One might think that we can learn of correlations between our decisions and past events only when we remember these events or are in possession of other reliable records of them, but to the extent that our memories or records are reliable, they screen off the past experiences from our present decisions. That is, when the second condition is satisfied, the first condition fails. By contrast, when $P(B/M(t)$ & $D(t)) \neq P(B/M(t))$, we cannot rely on any records to come to know the correlation between our decisions and a past event B. That is, when the first condition is satisfied, the second condition fails.

But this objection can be answered. We can learn of correlations between certain kinds of decisions and past events when we do have memories of the past events in question and then use that knowledge inductively to learn something about the past in cases where we make similar decisions but where the relevant memories are absent. This is not much different from how we come to believe reasonably that our decisions are correlated with future events – by learning inductively from experiences of past correlations.

Here is an example of this. While playing a piano piece that I know well, I am unsure whether I am currently playing a part of the piece that is repeated in the score for the first or the second time. I decide to play the second ending rather than repeating the part. Many of the notes I play, of course, I play without choosing or deciding to play them. But in the case I am imagining the question what notes to play next has arisen, and I consciously choose to play the second ending. Since I have learned from experience that when I play a piece I know well my decisions to play certain notes are good evidence for where I am in the piece, my present decision not to repeat the part constitutes good evidence for a certain past event – my having already played the part in question once. We can even imagine that I have a vague and unreliable memory of having already played the part. My decision to play the second ending, then, can constitute additional evidence for the reliability of my memory. In general, Loewer's first and second conditions are jointly satisfied in cases in which (i) we have good (inductive) reasons for treating our decisions as providing us with information about our past or past events in the world and (ii) the past events in question have left no or only very few and not fully reliable traces in the present.

As a third condition, Loewer requires that we have control over an event B only if it is part of the content of our decision that B occur. This third condition is not satisfied for events B in the past of the decision, since we do not (normally) take ourselves to have control over the past. But this last condition strikes me as too strong. We take ourselves to have control over · events that are consequences of our decision, even when the content of our decision is not that these events occur. For example, I may have the desire to arrive at the office by 9 a.m. and I have good reasons to believe that my arrival time is reliably correlated with the time when I leave my home. Then my decision to leave at a certain time provides me with a means of controlling when I arrive, even though my decision is, say, a decision to leave at 8 a.m. rather than a decision to arrive by 9 a.m. It seems that we can have control that p be so by decision, even when our decision is not a decision that p be so. Thus, two of Loewer's conditions on control by decision can be jointly satisfied by past events, while the third condition should be rejected on independent grounds.

Albert's and Loewer's aim is to offer an account of how it is that we possess a time-asymmetric concept of causal influence or control by showing that such a concept tracks certain non-causal features of the world given by fundamental physics, which can ground a counterfactual asymmetry that in turn can ground causal judgments. Commonsense causal claims are by and large concerned with relatively small, "human-sized" macro events of the kind that could be the result of human interventions. Arguably, then, any account of how we come to possess an asymmetric concept of cause need only be able to reproduce the asymmetry as far as causal claims within this domain are concerned.[11] It is, however, part of our notion of causal relations among "human-sized" macro events that the temporal asymmetry of causation is strict in the sense that, in *all* paradigmatic or standard circumstances, the relation of causation is future directed, and that in such cases there is absolutely *no* backward causation. We believe

---

[11] Thus, Albert rightly argues that the fact that a universe in the shape of Bozo the clown would have to have had a very different past from ours would not undermine his entropy-account of causation (see Albert 2000, 130, fn. 21). Even if Albert's account had the consequence that there is backward causal dependence in this case, this will pose no problem for his account, since this is not the kind of case that could have played a role in our acquisition of causal concepts.

that our interventions can have an effect on the future development of the world, and we also believe that our interventions can have *absolutely no* effect on the past. A successful statistical account would have to be able to account for this feature of our concept. This does not mean that it has to be a consequence of the account that there is no backward causal influence. Since Albert and Loewer's statistical account ultimately appeals to certain probabilistic relations that they take to be derivable from statistical mechanics, they may have the consequence that the causal asymmetry is not strict. Nevertheless, the account has to be able to explain why *we take* the asymmetry to be strict.[12] Thus, similar to derivations of the second law from statistical consideration, it would have to be shown that in paradigmatically causal contexts, exceptions to the asymmetry are extremely rare and improbable.

I have argued in the last three sections that Albert and Loewer's statistical account cannot underwrite a counterfactual asymmetry that is strict enough to ground our causal judgments. A crucial step in Albert and Loewer's account is an asymmetry of records: the claim that there are records of the past but not the future. In the next section I want to show that there is indeed such an asymmetry, but that it follows directly from the initial independence assumption and does not require the past hypothesis – that is, the assumption of a low-entropy past – as additional premise. This should come as no surprise, since recording interactions are causal interactions. Thus, just as an initial independence assumption and the causal asymmetry are intimately linked, so is the independence assumption and an asymmetry of records. But there is no need here to appeal to an asymmetry of counterfactuals. Rather, causal time-asymmetric counterfactuals can be introduced later, with their semantics given by Pearl-type structural models.

6. The asymmetry of records and the assumption of initial randomness

In this section I will show how an asymmetry of records – that is, that there are records of past events but not future events – can be derived from an assumption of initial randomness.[13] Let capital letters stand both for

---

[12] I think it may even be compatible with our commonsense notion of causation that there could be arcane physical circumstances in which there is backward causation. The point I am making here is that we take the asymmetry to be strict as far as commonsense, "billiard-ball-like" circumstances are concerned.

[13] Paul Horwich gives an argument to that effect (Horwich 1987), but Horwich's argument is flawed (see Healey 1991), and some of the argument's premises appear to be unmotivated.

variables and for the quantities or systems represented by these variables and small letters stand both for the values these variables can take and for the particular state represented by these values. R characterizes a record state. The value r^* is a putative record of the fact that the variable S characterizing the system of interest was in an earlier state s^*. R will in general have other causes aside from S. Here I want to distinguish causes of two different types.

First, there are boundary conditions B^k that must take on certain values b^k in order for R to provide a reliable record of the system S. In what follows I will leave these boundary conditions implicit. Conditionalizing the probabilities on appropriate values of B^k will make the expressions that follow look more cumbersome but will not affect the derivation. Whether R provides a reliable record of S may also depend on whether the recording device was in an appropriate "ready state" r_0 prior to the measurement. I will for now leave this assumption implicit as well (but will return to it later).

Second, there are n potential "external" causes, U^i. The variables U^i characterize both possible *macroscopic* states that can potentially influence what value S will take and *microscopic* variables characterizing the recording device or the boundary conditions B, which, if they were to take on certain special values, may threaten to undermine the correlation between the macro values of S and R. We will assume that the values u^i_j are distributed randomly, that is,

$$Pr\left(u_s^1 \wedge u_s^2 \wedge \cdots \wedge u_s^n\right) = \Pi_i Pr\left(u_s^i\right). \tag{1}$$

Clearly not every system can function as a recording device. One requirement for a good recording device is that its record states ought to be relatively stable and non-mixing for very long times. A second requirement, which will play a role in the argument that follows, is that the interaction between the system recorded and the measurement apparatus needs to be well shielded in a way that makes the state of the recording device relatively insensitive to the value of the external factors U^i_j. That such interactions exist does not follow from the initial randomness assumption, but rather depends on other physical facts concerning the system and interactions at issue. The shielding condition can be stated as demanding that (conditional on appropriate boundary conditions B and the system having been in its ready state r_0) conditional on almost all sets of values of the external variables U^i, the probability both of false positives and of false negatives will be low, that is: (i) the probability that R will have value r^* when S does not have value s^* is low; and also (ii) the probability that R will not have value r^* when S does have value s^* is low. That is, we will assume that only if the U^i all take on special values u^i_s will R, with high probability, be in the record state r^* even when S did not have value s^*. And equivalently, the shielding condition implies that there is only a special set of values u^i_t for which the recording device produces false negatives. I will here only focus on condition (i) and the possibility of false positives and assume that the recording interaction is perfectly shielded against false negatives. The argument in the case of false negatives proceeds strictly analogously to the argument that follows.

Are there recording interactions that are in fact strictly shielded – that is, are there interactions for which false positives or negatives are nomologically impossible, no matter what the values of the U^i_j are? The answer is "no." First, in practice no system can be *strictly* shielded against all external macroscopic influence; and second, it is impossible *strictly* to exclude what Carl Hoefer (2004) has called the "threat from within" – the possibility that the micro state of the system is highly abnormal in a way that prevents the system's macro evolution to be as expected.

The assumption that the probability of false positives, conditional on most values of the external factors U^i_j is low can be expressed as

$$Pr\left(r^* | \neg s^* \,\&\, \neg \forall i, j \left(u^i_j = u^i_s\right)\right) \approx 0. \tag{2}$$

One can now show that it follows from (1) and (2) that R is a reliable recording device. It follows from the probability calculus that

$$Pr(r^*) = Pr(r^*|s)\,Pr(s^*) + Pr(r^*|\neg s^*)\,Pr(\neg s^*). \tag{3}$$

Expanding the second term further results in

$$
\begin{aligned}
Pr(r^*|\neg s^*)&Pr(\neg s^*) \\
&= Pr\left(r^*|\neg s^* \,\&\, \neg \forall i, j\left(u^i_j = u^i_s\right)\right)Pr(\neg s^*)Pr\left(\neg \forall i, j\left(u^i_j = u^i_s\right)\right) \\
&+ Pr\left(r^*|s^* \,\&\, \forall i, j\left(u^i_j = u^i_s\right)\right)Pr(\neg s^*)Pr\left(\forall i, j\left(u^i_j = u^i_s\right)\right).
\end{aligned}
\tag{4}
$$

It then follows from (2) that the first term on the RHS of (4) will be very small and from (1) that the second term on the RHS will be very small as well, since

$$Pr\left(\forall i, j\left(u^i_j = u^i_s\right)\right) \approx 0. \tag{5}$$

Thus, from (2) and (5),

$$Pr(r^*|\neg s^*)Pr(\neg s^*) \approx 0. \tag{6}$$

And from (3) and (6),

$$Pr(r^*) \approx Pr(r^*|s^*)Pr(s^*). \tag{7}$$

Finally, from Bayes's theorem it follows that

$$Pr(s^*|r^*) = Pr(r^*|s^*)Pr(s^*)/Pr(r^*) \approx 1. \tag{8}$$

That is, it follows from the randomness assumption together with the assumption that R qualifies as a good recording device that our records of the past state of the system S are reliable.

We can now see why there are no (or at least many fewer) records of the future. A "recording" device of the future state of a system would be a device that has a relatively stable state r' that occurs with high probability whenever the recorded system at some *later* time is in a state s'. Both the stability and the shielding assumption can be satisfied by potential future records. But what blocks a derivation analogous to the preceding one to the conclusion that $Pr(s'|r) \approx 1$ is that we cannot assume future randomness. Rather, it follows from the initial randomness assumption together with the fact that different systems interact that there will be future correlations, sometimes merely hidden or microscopic, that can result in fake "pre-records."

To illustrate this point, consider the following example of a putative "pre-record," due to Lawrence Sklar. A fighter plane releases a bomb over a city. The release of the bomb, Sklar claims, satisfies the condition of being a pre-record of the destruction of the city. But I do not think that Sklar's claim is correct. For the city might not be destroyed despite the existence of the "pre-record." For example, if the city were protected by a missile defense system that was about to launch a missile, which destroyed the bomb in mid-flight, then the city would not be destroyed despite the release of the bomb. One might think that we can exclude this possibility or others like it by considering the state of the world in some larger region surrounding the bomb and positing that this state is such that it follows nomologically that the bomb will indeed hit the ground. But then the inference toward the future is no longer one relying merely on a localized "pre-record" but rather is an inference from the laws plus suitable initial or boundary conditions. Merely based on the existence of the pre-record, the presence of a missile defense system cannot be excluded. Now consider the state of the world at the time at which the bomb would have struck, had it not been for the missile. If the bomb was destroyed by a missile, then there will be scattered, perhaps microscopic, traces of the midair explosion – traces that are highly correlated in such a manner that they together can "account for" the existence of the misleading pre-record of the city's destruction. Thus, in this case we cannot impose a condition of final randomness that would render such correlations among later "external" states highly improbable.

As I said earlier, the result derived in this section should not be surprising. Records exhibit a common-cause structure: multiple records of the state of a system at time t and the state of the system at times later than t are all joint effects of the system's state at t, satisfying a screening-off condition. Thus, just as an assumption of initial randomness implies the Markov condition, it also can underwrite a record condition.

7. Price's subjectivist account

According to Albert and Loewer's account, the probability postulate together with the past hypothesis grounds an asymmetry of certain counterfactuals (such as Loewer's "decision counterfactuals"), which in turn account for our employment of time-asymmetric causal notions. The central problem for their account, I argued, is that Albert and Loewer's counterfactuals do not exhibit a temporal asymmetry that is strict enough to underwrite a strict causal asymmetry. A somewhat different strategy is pursued by subjectivist account of causation according to which the fact that we describe the world in asymmetric causal terms has its origins in a more fundamental asymmetry of deliberation. Accounts along these lines have been proposed by Healey (1983) and Price (1997). I here want to focus on the recent defense of such an account developed in Price and Weslake (2009).

Price and Weslake's starting point is an evidentiary asymmetry of deliberation consisting in the fact that we deliberate with future ends and not past ends in mind. They argue, first, that our deliberations exhibit a temporal asymmetry that underwrites an asymmetry of influence: "we can't use evidence as a 'causal handle' to influence the earlier states of affairs for which it provides evidence" (433). This asymmetry, they claim, can then be extended to cover causal claims in general, since it provides us with a time-asymmetric perspective on the world, which underwrites a time-asymmetric prescription as to how to evaluate counterfactual interventions into systems not involving human agents: "it is our perspective as *deliberators* that underpins the distinction between cause and effect," they emphasize. From this it follows, Price and Weslake maintain, that causal and counterfactual claims can be treated in terms of interventionist accounts of causation, such as the ones developed by Pearl or Woodward. But unlike Pearl or Woodward, Price and Weslake do not introduce asymmetric causal relations as basic, but rather take them to be a projection of a more fundamental asymmetry of deliberation.

Price and Weslake argue, with F. P. Ramsey, that *during the process of deliberation* we have to think of our actions as being uncaused: "my present action is an ultimate and the only ultimate contingency" (Ramsey quoted in Price and Weslake 2009). To the extent that we conceive of our decisions as free, we cannot consider them as being correlated with any other events except for their effects. But what accounts for the deliberative asymmetry, if it does indeed exist? Why do we deliberate with future rather than past ends in mind? The fact that we are beings with a certain temporal bias that manifests itself in the deliberative asymmetry, Price and Weslake maintain, is ultimately explicable in terms of the statistical asymmetry underlying thermodynamic phenomena. But the latter asymmetry, they assume, cannot itself directly ground the asymmetry of causation; it can only do so via a detour through the deliberative asymmetry characteristic of agents like us. The existence of a local thermodynamic gradient results in the deliberative asymmetry, which then locally fixes a direction of causation. We then "spread" this local asymmetry of causation "over the objects," to adopt Hume's phrase, and extend it into a global asymmetry of causation. This is possible as long as the universe we live in is temporally orientable.

Price and Weslake take the deliberative asymmetry to be more fundamental than the causal asymmetry, while according to a causal theory of deliberation, Price and Weslake's subjectivist theory inherits the deliberative arrow from the causal arrow and then spreads that arrow over the universe as a whole. How can we adjudicate between these two accounts? To the extent that our universe is causally "normal," the two proposals will agree not merely locally but also globally on the direction of causation, but a way to distinguish the two proposals is to examine what they would say about hypothetical universes in which there seem to be oppositely directed causal arrows.[14]

Price considers the possibility that the universe may exhibit a global time symmetry and not only did evolve from an initial low-entropy macro state together with a condition of initial randomness, but also will evolve toward a future low-entropy state that satisfies a condition of final randomness (Price 1997). If we posit that the age of such a universe is more than twice its relaxation time, then the two periods of entropy increase and decrease are separated by a time period in which the universe is in a maximum-entropy

---

[14] Newton-Smith (1983) makes an argument similar to the one that follows against the subjectivist account proposed in Healey (1983).

equilibrium state. The thermodynamic arrow points in opposite directions during the two temporal halves of the universe. Which of these two directions is the future direction? Price's suggestion is that beings living in either half of the universe would disagree on a global temporal arrow and would identify the direction of entropy increase within their respective local environment as the future direction. Let us also assume that both epochs can be characterized in terms of a causal asymmetry. In which direction does the causal arrow point during the two epochs?

In fact, it is not obvious what the answer to this question ought to be in Price and Weslake's account. On the one hand, they maintain that the direction of the causal arrow globally is simply the same as that of our local arrow of deliberation. Yet on the other hand they suggest that one can extend the deliberative asymmetry to causal claims not involving human agency by using interventionist accounts of causation: "Ideally, the subjectivist will want to step into the interventionists' shoes – all the more so, now that Pearl, Woodward and others have shown us how far those shoes may take us!" (Price and Weslake 2009, 438). But for regions of the universe with opposite entropic arrows, the two prescriptions give conflicting results. According to Price and Weslake's first prescription, the arrow of causation, as *projected by us*, globally points in the same direction as our deliberative arrow does locally. Thus, *for us* the causal arrow during the entropically reversed epoch points from states with higher entropy to states with lower entropy. But from our temporal perspective, the entropy-decreasing epoch is one in which there exist delicate correlations among micro states of physical systems that are just so that systems will "miraculously" evolve from states that are macroscopically more random to macroscopically more highly ordered states. At the same time, micro states will evolve from highly correlated states to states that are microscopically random. This can be best seen by considering the fact that the entropy-decreasing epoch is thermodynamically simply the time-reverse of the thermodynamically normal behavior with which we are familiar. During our epoch, ordered macro states evolve into less ordered macro states, while systems begin in micro states that are "typical" or random, given the corresponding macro state of a system, and evolve into micro states that are highly correlated, having evolved from a "special" low-entropy initial macro state. That is, during the entropy-decreasing epoch we would take higher entropy states exhibiting microscopic correlations to be the cause of later lower entropy states, which are microscopically more random.

Contrast this with the second, interventionist prescription (as formalized by Pearl and others). According to that prescription the randomness assumption holds in the causal *past* of the models: exogenous variables are distributed randomly. Thus, according to the two different prescriptions, the causal arrow will point in opposite directions during the two epochs of our hypothetical universe. The interventionist models satisfy the causal Markov condition and, therefore, a screening-off condition. The causal models resulting from Price and Weslake's first prescription, by contrast, are non-Markovian. In fact, the prescription implies not only an isolated failure of the Markov condition but a general failure of the condition during the entire (from our perspective) anti-entropic epoch of the universe. Yet without the Markov condition and the ability to engage in common-cause reasoning, the whole point of representing the world in terms of causal structures seems lost. Indeed, as Pearl maintains, it is "not clear how one would predict the effect of interventions from such a [non-Markovian] model, save for explicitly listing the effect of every conceivable intervention in advance" (Pearl 2009, 61).

Finally, if Price and Weslake allow the two causal arrows to come apart in some regions of the universe, the question arises as to why we should identify the two arrows in our own epoch of the universe. Price and Weslake, thus, seems to face the following dilemma: They can follow Pearl and others in positing as default (but in isolated special cases perhaps defeasible) constraint on causal models that causal models are Markovian and hence that disturbances in the causal past are distributed randomly. Then the causal and entropic arrows will be aligned during both epochs of a symmetric universe, pointing in opposite directions during the two epochs, but Price and Weslake's "detour" through our own perspective of deliberation in assigning a global causal arrow does no work. Or they can take the direction of causation everywhere to be that given by the local arrow of deliberation and action. But this has the consequence that we are now faced with two distinct arrows that might be called "causal": the perspectival causal arrow on which Price and Weslake focus and the interventionist inferential arrow. But since we are now allowing that the two arrows can come apart, we need an explanation why the arrow of deliberation lines up with the entropic and interventionist causal arrow locally.

I have an additional, more general worry about Price and Weslake's account. The starting point of the account is an asymmetry of deliberation, which arises, they argue, even if we think of deliberation in purely "epistemic, evidential or 'pre-causal' terms" (Price and Weslake 2009). The asymmetry is said to consist of our believing a certain material conditional – that if we perform an action *A*, then an outcome *O* will occur – where *A* precedes O. But if the connection between A and O is construed purely evidentially, and not already causally, it is far from clear that we only make inferences to later outcomes from our decisions or actions and not also to earlier states of the world. I presented an example of an inference from a decision to an earlier state above, and Holton discusses others, arguing that one important role of decisions is in fact to provide not explicitly represented information about the world to an expert actor (Holton 2006). Now, Price and Weslake argue that *during* the moment of deliberation we must treat our decision as free and hence as not being evidentially informative about past events. But, first, this assumption may be questioned, and second, it is unclear what epistemically privileges the *moment of decision* over other moments when we reflect on our decisions and on what these decisions might tell us about who we are and about our pasts. Now, one might think the moment of decision and action has a special status in an account of the causal asymmetry, because it is precisely that moment that provides us with a direct and unmediated experience of causation. But that line of argument is, of course, not open to Price and Weslake, who want to take a non-causal asymmetry as their starting point. And evidentially, our decisions can be a source of information about the future as well as the past.

8. Conclusion

Both Albert and Loewer's thermodynamic account of the causal asymmetry and Price's agency account stress the tight link between an initial randomness assumption and the causal asymmetry. Both accounts try to show how this assumption can feature in a reductive account of the causal asymmetry – in connection with the asymmetry of thermodynamics in Albert and Loewer's case and via an asymmetry of deliberation in Price's case. Yet I have argued here that neither Loewer and Albert's attempt to ground the causal asymmetry in a statistical asymmetry through a Lewis-style counterfactual account of causation nor Price's attempt to ground the causal asymmetry in an asymmetry of deliberation are successful.

Both reductive accounts aim to locate the causal asymmetry at some remove from the putatively metaphysically fundamental level, the Humean mosaic of particular non-modal matters of fact. Albert and Loewer's account sees causal notions as being grounded in a Lewis-style account of counterfactuals and laws, which in turn are reducible to the Humean mosaic. Price's account takes the route through an agent-centered asymmetry, which presumably also can be further reduced to non-modal physical asymmetries characterizing the mosaic. One lesson I want to draw from the problems of these accounts and from my derivation of an asymmetry of records earlier is this: if one wanted to offer a reductive account of the causal asymmetry at all, then it would be more promising to try ground the causal asymmetry directly in the initial randomness or independence assumption, rather than taking the additional detours Albert, Loewer, and Price propose.

CHAPTER 9

Conclusion

In the preceding chapters I argued that time-asymmetric causal relations *can* – and, as a matter of fact, *do* – play an important role in physics. I argued, first, that several influential arguments that aim to draw a distinction between representation in physics and representations in the special sciences fail. Representations in physics, just like representation in the special sciences, are partial and coarse-grained and involve a distinction between model and background conditions. Moreover, models in physics are not incompatible with what are arguably two core properties of causal structures: causal structures are asymmetric and, in particular, are time-asymmetric; and causal structures underwrite and in turn can be justified by interventionist reasoning. Modeling in physics, I concluded, is no less hospitable to causal reasoning and causal structures than is modeling in the special sciences.

I argued, second, that causal reasoning does, as a matter of fact, occupy an important place in how we represent the world within the context of established theories of physics. According to one influential account of scientific representation in physics, a theory's laws together with specific initial or final conditions define individual representational structures, and this exhausts the representational resources physics has at its disposal. Indeed, if initial conditions and laws are fully specified, then there seems to be no room for causal assumptions to play a substantive role. Against this view I argued that in representing actual physical systems, the antecedent of the preceding claim rarely is satisfied. If pure initial- or final-value problems provided us with the only tool at our disposal for making inferences from the state of a system at one time to its state at any other times, then we could know precious little about the physical world.[1] Very rarely, if ever, do we have the empirical data to specify the complete state of a system on a relevant

---

[1] David Albert makes a similar point in Albert (2000), in an argument not for the need of causal assumptions but for positing the past hypothesis. For a critical discussion of Albert's account of the role of records, see Frisch (2007).

initial-value surface. Thus, very rarely, if ever, can we use the machinery of pure initial- or final-value problems to make inferences from one time to another. This is where causal assumptions enter. The predominant – and indeed perhaps the only – way to extend our epistemic reach, when we lack complete initial data, is with the help of time-asymmetric causal structures. Thus, the representational resources employed in physics have to be richer than the standard account allows.

Formally, the kind of causal reasoning I discussed can be reconstructed in terms of a structural account of causation, such as that developed by Judea Pearl. In Pearl's account, a *causal model* consists of a directed acyclic graph over a set of variables $\{X, Y, \ldots\}$; structural equations $x_i = f_i(pa_i, u_i)$, which specify the value of each variable in terms of the value of the variable's causal parents pa_i and random exogenous disturbances u_i; and a probability distribution $P(u_i)$ over the values u_i of the exogenous variables U_i. As Pearl introduces the notion, causal models are discrete structures. We can think of introducing causal structures in physics as superimposing a time-asymmetric structure over a discretized version of a theory's dynamical model. For example, in the case of observations of the light emitted by a star, the variables over which the model is defined are variables representing the set of observed field values, the state of the star, and weak free incoming fields. The state of the field in between the emission and observation events is not explicitly represented in the model. The structural equations relating the values of the variables, I argued, are given by causal Green's functions, which are determined by a theory's dynamical equations. The Green's function formalism also allows us to introduce models with a continuum of variables representing the state of a system at each spacetime point in the region occupied by the system. What the causal relation is between each pair of spacetime points is given by the causal Green's function, which determines how disturbances propagate through the system.

In Chapters 6 and 7 I examined two cases of theorizing in physics in which time-asymmetric causal assumptions play an important role: linear response theory and explanations of the wave or radiation asymmetry. In the first case – that of linear response theory – there is a unique Green's function associated with the problem that can readily be interpreted causally. In the case of the wave equation, the Green's function is not unique, yet the retarded or causal Green's function is privileged over the advanced Green's function due to the asymmetry between prevailing initial and final conditions. Causal models constructed with the help of the causal or retarded Green's function satisfy the causal Markov condition, whereas putative causal models constructed from the advanced Green's function are not Markovian.

I argued that representation in physics has much more in common with representation in other sciences than is often assumed. Many of the putatively "human-faced" features that Jim Woodward and Hartry Field, among others, attribute to causal representations are characteristic of model building in physics as well. One might object, however, that neither these arguments nor my arguments appealing to circumstances in which a system's full initial conditions or full micro state are unknown can show that causal relations play a role in how (in some sense) *fundamental* physics represents the world. My discussion, one might claim, merely shows that modeling in *non-fundamental* or *applied* physics shares crucial features with causal modeling elsewhere. Moreover, my claim that causal notions allow us to draw inferences from incomplete evidence only further serves to emphasize the stark difference between causal and nomological constraints in physics: a Laplacian demon who had access to the complete state on an initial-value surface would have no need for causal assumptions but could not do without the dynamical laws. Thus, although causal notions might satisfy what Woodward singles out as the only acceptable standard of legitimacy, usefulness, one might think that causal notions are useful only as a preliminary epistemic "crutch." I want to offer several replies to this objection.

In *How the Laws of Physics Lie* (Cartwright 1983), Nancy Cartwright introduces a distinction between two different kinds of model of physical theories: On the one hand there are models used to represent actual physical phenomena, such as models of the LHC, which I discussed in earlier chapters. Cartwright calls such models "representational models." On the other hand there are models that are constructed to satisfy a theory's laws exactly; the laws, that is, are strictly true of these models. These models are, in some informal sense, model-theoretic models of the theory's laws. Some such models represent extremely simple "possible worlds," and these are the kinds of models physics students study and investigate in problem sets. Cartwright calls these "theoretical models." Yet in appealing to the make-believe construction of a Laplacian demon, we are also positing highly complex models of which the laws as strictly true and which represent worlds as complex as the actual world – incredibly complex models imagined (but only imagined!) to be specified in complete and microscopic detail. It is this last type of model that, according to the causal skeptic, is inhospitable to causal notions.

What should we make of this appeal to these hypothesized complex models? First, if we accept, as I have argued in Chapter 2 we should, that a model is a representation of a phenomenon only if it is *used* as a model of that phenomenon, then the complete possible worlds postulated by the philosopher to satisfy a theory's fundamental equations cannot be representations at all, since they are too complicated to be grasped – let alone used – by us. Moreover, it is unclear how much stock we should put in an appeal to impossibly complex models putatively representing worlds as complex as ours. Indeed, it seems to me that the causal skeptic has been seduced by a powerful and widely accepted image of the content of a physical theory according to which what an established physical theory tells us about the world is fully and completely embodied in the hypothesized immensely complex solutions to the fundamental equations, which are postulated to capture real-world phenomena in complete detail. Yet to assume that the content of a physical theory is exclusively given by these complete structures, which we do not and cannot have "in our hands," rather than by the models that physicists *actually use* to represent actual physical systems seems to me to get things exactly backward. We use theories to represent the world, and what a theory tells us about the world is given by those structures that we use to represent actual phenomena. If we want to find out what a theory tells us about the world, we should look to the representational models constructed with its help.

Second, when we *confirm* a theory's basic equations, we do this by confirming representational models constructed with their help. But, as I have argued, causal assumptions play an indispensable role in the construction of at least certain types of representational models. That is, the models we actually confirm are, in the first instance, not merely dynamical models, since we generally do not have the kind of data needed to directly test a specific dynamical model, but rather are what I want to call *causal dynamical models* that are constructed *both* with the help of dynamical equations *and* with the help of time-asymmetric causal assumptions (see Figure 9.1).

Consider as an example once again a model of the light emitted by a star. There are several different dynamical models, each associated with a different distribution of initial fields and sources, into which our observations of light points can be embedded. Our extremely limited observations leave the choice of dynamical model radically underdetermined. Yet if we supplement the dynamical model with causal structures, we can solve the underdetermination problem, if we demand that a causal model into which representations of the correlated observations of light points are embedded

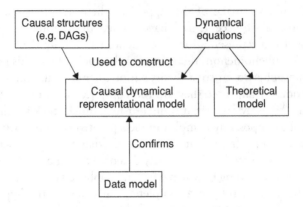

Figure 9.1 Confirming causal structures

satisfy the causal Markov condition. The causal dynamical model we use to represent the phenomenon consists of a causal structure – the emission of light by the star and its common effects, the light points observed on Earth – and structural equations, which are given by the causal Green's function of the wave equation.

Another way to put this point is this: Since we do not have access to the full final conditions needed to set up and solve a pure final-value problem to retrodict the emission of light by a star, we use the retarded Green's function alone to infer the state of the source from the radiation we observe. And what justifies our use of the retarded Green's function, instead of the advanced Green's function or of any linear combination of the two, is a causal inference from the existence of the correlations among our observations to a common cause.

If this is right, then what the causal skeptic has to assume is that, once we have confirmed a causal-dynamical representational model, confirmation flows further upward (as it were) from the causal dynamical model only to the theory's laws, and not to the causal assumptions made in constructing the model. Or, to put this in terms of Pearl's structural models, what the causal skeptic has to claim is that what we can confirm are the *structural equations* (given by the causal Green's function associated with the dynamical equations), but now read as *ordinary equations*, but that we cannot also confirm the *acyclic causal graph* associated with the model. It is unclear what the justification for this unequal treatment might be. More plausibly, confirming a representational model provides us with a prima facie reason to accept all the assumptions made in constructing the model: if a model, which we confirm through experiment and observation, is constructed with the help of causal assumptions, we confirm these assumptions as part of the representational resources we use in physics.

Third, it is far from clear what the epistemological or metaphysical relevance is of the conceit of a Laplacian demon, who is not in need of making causal assumptions. For one, the Laplacian conceit puts a severe and arguably unjustified constraint on the form dynamical laws can take. As I have discussed in Chapter 3, a well-posed pure initial-value problem exists only for hyperbolic differential equations and not for parabolic or elliptic equations, such as the Navier-Stokes equation in fluid dynamics or (depending on the Hamiltonian) the Schrödinger equation.[2]

Yet let us assume for the purposes of the argument that all the dynamical equations of all established and suitably fundamental theories of physics are hyperbolic. We can contrast the following two epistemic perspectives on the world, which occupy opposite ends in a range of possible perspectives. First, there is our limited and partial perspective, which requires us to make use of both causal assumptions and dynamical laws in representing physical phenomena. At the other extreme we might imagine the perspective of a truly omniscient being, a being like Albert and Loewer's God (see Chapter 8), who has direct knowledge of the entire Humean mosaic – the complete spatiotemporal mosaic of particular matters of fact – and whom we might ask for the best way of summarizing these facts. This being is in the possession of a truly "absolute conception" of the world, as Bernard Williams has called it (Williams 1978). Loewer's and Albert's God has a need neither for dynamical laws nor for causal assumptions – She can directly survey all matters of fact without the need to derive facts at one time from what She knows about facts obtaining at another time.

The challenge for the causal skeptic, as I see it, is to show why the conceit of a Laplacian demon, who occupies an intermediate epistemic perspective between us humans and Albert and Loewer's God, should have any methodological or metaphysical relevance. *We* need both dynamical laws and causal assumptions "to make our way about in the world"; *Albert and Loewer's God*, by contrast, needs neither. What then is the relevance of the imagined Laplacian demon's intermediate epistemic position?

One might try and argue that the demon's epistemic situation shows that it is at least in principle possible to do without causal assumption,

---

[2] A problem is well posed if a solution exists, the solution is unique and the solution depends continuously on the initial or boundary data (see, e.g., Snider 2006, 265).

but there is no equivalent "causalist" demon, who can derive the full Humean mosaic from partial knowledge of the mosaic and knowledge only of causal principles. That is, the conceit of the Laplacian demon points to an asymmetry between laws and causal assumptions. But as I have stated it, the premise of the argument is false: if we assume that a putative causalist demon is in possession not only of the *causal graphs* to generate his predictions but also of the *structural equations* associated with the causal structures – that is, if he is in possession of the causal Green's functions associated with the dynamical equations – then his derivational power is in fact equivalent to that of the Laplacian demon. Just as the Laplacian demon can derive the state of the mosaic in an arbitrary from appropriate initial conditions, the causalist demon can derive the state of the system from the knowledge of the Green's functions and appropriate boundary conditions. A solution to the dynamical equations in any bounded region can be decomposed into a conjunction of three terms: the first representing the dependence on the inhomogeneities, that is, the sources located within the bounded region; the second representing the dependence on the initial conditions; and the third representing the dependence on the boundary conditions.[3] All three terms can be derived from the causal Green's function. That is, the Green's function determines not only how the presence of sources affects the state of a system, but also how inhomogeneous (that is, non-zero) initial and boundary conditions propagate into a bounded spacetime region.

There is a further reply to the worry that I have only shown that causal structures play a role in applied physics and in cases where we possess only partial knowledge of the state of a system: causal structures play an explanatory role even in situations where we imagine ourselves to have full knowledge of a system's initial or final conditions. And as we will see, the explanatory role of causal structures may tip the scales in favor of taking time-asymmetric causal relations to be conceptually prior to the initial randomness or independence assumption.

Time-asymmetric causal structures reflect a deep explanatory asymmetry between past and future states. When we observe correlations among spatially distant events, we seek to explain the correlations in terms of earlier states that act as common causes rather than in terms of later states. Recall Earman's contrast between a broadcast and an anti-broadcast antenna, discussed in Chapter 7. Both antennas are associated with

---

[3] See Barton (1989), who derives this result, which he calls the "magic rule" for the diffusion equation and for the wave equation. See also Smith (2013).

correlations in the radiation field. But whereas in the case of the broadcast antenna these correlations can be satisfactorily explained by an appeal to the earlier action of the antenna, in the case of the wave collapsing into the anti-broadcast antenna the correlations strike us as "miraculous," as Earman himself stresses. If anything, the correlation between the later state of the antenna and the earlier state of the field seems to add to the sense of mystery. Representing the relation between fields and antennas in terms of time-asymmetric causal structures reflects this explanatory asymmetry: In the case of the broadcast antenna, the action of the antenna is the common cause of the correlated subsequent disturbances in the field. In the case of the anti-broadcast antenna, there is no common cause, and the correlations remain mysterious. Thus, we can explain – and render non-miraculous – correlations in terms of earlier common causes, but we apparently cannot explain correlations without prior common cause by appealing to later events.

As Maudlin has argued (Maudlin 2007, 131–4), just such an asymmetry between initial and final states also underlies an explanation of the thermodynamic asymmetry, which also supports a causal interpretation of the relation between earlier and later states. The universe began in a state that is *macroscopically atypical* (in that it was a state of extremely low entropy) but *microscopically random* or typical. The final state of the universe will be *macroscopically typical*, since it will be an equilibrium state, but it will be *microscopically atypical*, since it will be a state with just the right delicate correlations to evolve backward in time into the macroscopically atypical initial state. (Think of a gas that was confined to a small volume and then spreads out in a box: the final micro state of the gas will contain delicate correlations such that if we let the state of the gas evolve backward in time, the molecules all concentrate in the small volume in which the gas was initially.) We can explain the universe's thermodynamic evolution in terms of the dynamical laws and a microscopically typical (but macroscopically atypical) initial state. But, Maudlin argues, we cannot similarly explain the universe's evolution by appealing to its final state, since the way in which this state is microscopically atypical cannot be characterized independently of the universe's prior evolution. Indeed, as far as a system's thermodynamic behavior is concerned, micro states are always *atypical* as far as a system's backward temporal evolution is concerned. And this can be explained by the fact that earlier states cause or produce later states. As in the examples I discussed earlier, a randomness assumption holds for initial but not for final states, and correlations in the final micro state are explained by the state's prior evolution.

The temporal asymmetry of causal structures, thus, reflects what appears to be a deep asymmetry characterizing our explanatory practices, even when we assume the complete state of a system at a time to be known. As I have described the explanatory asymmetry so far, however, it may seem that we can equally as well account for it, if we assume the initial randomness assumption as basic. Recall the mutual dependence between the causal Markov condition and the assumption that exogenous variables are probabilistically uncorrelated. One might take the initial independence assumption to be itself a causal assumption, as Pearl does (see, e.g., Pearl 2011, 704) and take it to be a consequence of the assumption that exogenous variables have no common causes in their past, as Woodward and Hausman argue. Or one might hold that the probabilistic independence assumption underwrites the legitimacy of representing phenomena in terms of causal structures.

If, for example, we assumed a starkly "Humean" metaphysics, then both laws and causal assumptions are ultimately reducible to facts about the Humean mosaic: neither are fundamental, yet both play a useful role in summarizing facts about the mosaic. The laws capture regularities concerning how states dynamically evolve, while the causal structures allow us to capture a global asymmetry characterizing the mosaic – an asymmetry between prevailing initial and prevailing final conditions. Both dynamical laws and the causal asymmetry, according to the Humean picture, supervene on facts about the mosaic, but they capture different aspects of the mosaic: the dynamical laws summarize regularities exhibited by the mosaic, whereas causal structures are underwritten by the asymmetry between initial and final conditions.

Thus, both a Humean, who views the initial randomness assumption to be a property of the Humean mosaic, and someone who believes in metaphysically more robust relations of causal production can allow for causal notions to play an integral role in how physics represents the world. And both views satisfy Woodward's condition on a functional account of causation of being committed to a weak realism about causal relations: On both accounts "the difference between those relations that are merely correlational and those that are causal has its source 'out there' in the world" (see Chapter 1). For the Humean the source is a temporal asymmetry in the distribution of matters of fact, while for the non-Humean the source is the existence of time-asymmetric causal relations between events. Yet against both types of metaphysical views I suggested earlier that our realist commitment might be even thinner: there might not be a uniquely correct and interesting and context-independent answer to the question whether the initial independence assumption or the causal asymmetry is more fundamental, and hence we might not be able to identify a metaphysically fundamental or basic feature of the world that underlies the success of causal representations.

But the sense that correlations without prior cause are "miraculous" invites a further type of explanation request, which we need to distinguish from the kind of explanation so far and which seems to favor taking the causal asymmetry to be primary. So far we have focused on explanations for correlations among events. As far as such explanations are concerned, the causal Markov condition and the initial randomness assumption may be viewed as two interderivable properties of the causal structures employed in the explanations. The explanations will be successful exactly if the phenomena can be represented in terms of causal structures that satisfy the causal Markov condition and, hence, an initial independence assumption.

There is a second type of explanation-request, however – one concerning prevailing initial conditions – which can arguably break the stalemate. The dynamical laws allow us to calculate what the state of the world is at any other time, given the world at one time (or, more carefully, on one spacelike hypersurface). Yet the "mosaic" at which we arrive by stitching together states of the world at different times exhibits a striking overall asymmetry: the universe evolves from states that are microscopically random, uncorrelated, or "typical" to states that are microscopically highly atypical and correlated. Indeed, correlated initial states would strike us as "near-miraculous," as Earman says, while random or uncorrelated initial states strike us as normal and not in need of an explanation. Earman goes on to ask what might explain the absence of such "near miracles" and the asymmetry between prevailing initial and final conditions. That is, not only might one ask what can account for the presence of correlations within the mosaic by searching for a prior common cause, but one might also ask why the mosaic exhibits an overall and global asymmetry (at least as far as it presents itself to us). If the absence of "near-miraculously" correlated initial states is indeed something that may call for an explanation, then the causal asymmetry is conceptually prior in causal structures to the initial randomness assumption. Earman proposes as explanation for this asymmetry an "improbability in the coordinated behavior of incoming source free radiation from different directions in space" (see Chapter 7). But why should, from a completely time-symmetric perspective, coordinated source-free incoming radiation be any less probable than coordinated source-free outgoing radiation? The initial randomness assumption amounts to nothing more but a restatement of the asymmetry. By contrast, the causal asymmetry can explain it: the actual world contains broadcast antennas but not anti-broadcast antennas, because earlier states of a system cause later states, and we would not expect to find the kind of correlations associated with the anti-broadcast antenna without a prior cause.

Thus, if we assume that the asymmetry between prevailing initial and final conditions is in need of an explanation, this seems to favor a non-Humean view that takes causal relations to be more fundamental than the initial randomness assumption. For the non-Humean, the dynamical laws do genuine explanatory work – they explain how one state evolves into another – but so do causal assumptions, which explain the asymmetry between prevailing initial and final conditions. Moreover, according to the non-Humean, it is a consequence of the causal asymmetry that explanations appealing to laws are generally time-asymmetric, and we can explain the state of a system at one time t by appealing to how it evolved from a past state of the system, but not by appealing to the system's future evolution at times after t.

I believe it is telling that even skeptics of causal notions in physics, such as Earman and Norton, believe that there is an explanatory asymmetry between a broadcast antenna and an anti-broadcast antenna – that is, that there is an explanatory asymmetry between prevailing initial and final conditions. The asymmetry is so deeply engrained in our explanatory practices that even causal skeptics apparently take it for granted. Yet this observation does not yet settle the issue, since one might want to reject the explanatory asymmetry, as pervasive as it may be, as nothing more but the consequence of a human "prejudice" shaped by the fact that we are embedded in an environment that, as it happens, exhibits a certain temporal asymmetry.

However that may be, causal structures are an important part of the toolkit we use to represent physical phenomena. This conclusion is compatible with a broad range of attitudes toward these structures, from a thoroughgoing metaphysical realism to more instrumentalist attitudes. The claim I have defended here is only that causal assumptions ought to be treated on a par with other representational resources in physics and in science more generally.

In his book on probabilistic theories of causation, Patrick Suppes characterizes contemporary physics this way: "What we are able to get a grip on is a variety of heterogeneous, partial relationships. In the rough and ready sense of ordinary experience, these partial relationships often express causal relations, and it is only natural to talk about causes in very much the same way that we do in ordinary experience" (Suppes 1970, 6). According to Hitchcock's reading of Suppes's discussion, Suppes is arguing that there are preliminary stages in the treatment of phenomena in physics – stages at which these phenomena "still await systematic treatment" – when offering a qualitative causal characterization of the phenomena in question may be legitimate. But as our understanding of the phenomena progresses, causal talk eventually ought to be eliminated in favor of a more systematic quantitative treatment. Hitchcock proposes a principle to capture this view, which I quoted earlier: "There are advanced stages in the study of certain phenomena when it becomes appropriate to eliminate causal talk in favor of mathematical relationships (or other more precise characterizations)" (Hitchcock 2007, 56).

I do not here want to engage in a careful exegesis of Suppes's views, but it seems to me that one could also interpret Suppes's remarks quite differently from the way in which Hitchcock does. According to Hitchcock's reading, it is a peculiar feature of causal characterizations of phenomena that they are partial, imprecise, merely qualitative, and preliminary. According to the alternative reading I want to propose here, Suppes is drawing attention primarily to a general feature that all representations in physics share with those in the higher sciences and in common sense: their *partiality*. Classical physics may have seduced us with the illusion that physics would ultimately provide us with complete and exact representations of the phenomena, but this was never more than just that – an illusion. Within the partial and heterogeneous representations of the phenomena, which physics can actually offer, causal representations play a legitimate and important role. Recall our discussion of the LHC. The models of the proton beam, the accelerator, the particle detectors, and of various external influences are partial and are built with an extremely heterogeneous range of modeling tools, ranging from elementary particle physics and classical electrodynamics to phenomenological treatments and the mathematics of control systems. The models treat various types of interventions into the beams and the construction of some of the models involves what are explicitly identified as time-asymmetric causal assumptions. Indeed, it is precisely the partiality of representations, limits on the evidence available to us, and the treatment of various temporal asymmetries that provide an argument for the need for causal representations even in physics.

Bibliography

Albert, David Z. 2000. *Time and Chance*. Cambridge, MA: Harvard University Press.

2012. Physics and Chance. In *Probability in Physics*, edited by Yemima Ben-Menahem and Meir Hemmo, 17–40. The Frontiers Collection. Berlin: Springer.

Arntzenius, Frank. 1992. The Common Cause Principle. *PSA: Proceedings of the Biennial Meeting of the Philosophy of Science Association* 1992: 227–37.

Arntzenius, Frank. 2010. Reichenbach's Common Cause Principle. In *The Stanford Encyclopedia of Philosophy*, edited by Edward N. Zalta. http://plato.stanford .edu/archives/fall2010/entries/physics-Rpcc.

Atkinson, David. 2006. Does Quantum Electrodynamics Have an Arrow of Time? *Studies in History and Philosophy of Science Part B* 37 (3): 528–41.

Baez, John. 2011. *The End of the Universe*. http://math.ucr.edu/home/baez/end. html.

Bartels, Andreas. 2005. *Strukturale Repräsentation*. Paderborn: Mentis-Verl.

2006. Defending the Structural Concept of Representation. *Theoria* 21 (55): 7–19.

Barton, Gabriel. 1989. *Elements of Green's Functions and Propagation: Potentials, Diffusion, and Waves*. New York: Clarendon Press; Oxford University Press.

Bohr, Niels. 1948. On the Notions of Causality and Complementarity. *Dialectica* 2: 312–19.

'Bogen, James, and Woodward.' 1988. Saving the Phenomena. *The Philosophical Review* 97 (3): 303–52.

Born, Max, and Emil Wolf. 1999. *Principles of Optics: Electromagnetic Theory of Propagation, Interference and Diffraction of Light*. New York: Cambridge University Press.

Bueno, Otávio, and Steven French. 2011. How Theories Represent. *British Journal for the Philosophy of Science* 62 (4): 857–94.

Campbell, John. 2006. An Interventionist Approach to Causation in Psychology. In *Causal Learning: Psychology, Philosophy and Computation*. New York: Oxford University Press.

Cartwright, Nancy. 1979. Causal Laws and Effective Strategies. *Noûs* 13 (4) (November 1): 419–37.

1983. *How the Laws of Physics Lie*. New York: Oxford University Press.

1989. *Nature's Capacities and Their Measurement*. New York: Clarendon Press; Oxford University Press.

1993. In Defence of "This Worldly" Causality: Comments on van Fraassen's Laws and Symmetry. *Philosophy and Phenomenological Research* 53(2): 423–29.

1999. *The Dappled World: A Study of the Boundaries of Science*. New York: Cambridge University Press.

2001. What Is Wrong With Bayes Nets? *The Monist* 84 (2): 242–64.

2010. Natural Laws and the Closure of Physics. In *Visions of Discovery: New Light on Physics, Cosmology, and Consciousness*, edited by Raymond Y. Chiao, Marvin L. Cohen, Anthony J. Leggett, William D. Phillips, and Charles L. Harper Jr., 612–23. Cambridge, UK: Cambridge University Press.

Cohen, I. B. 1978. *Isaac Newton's Papers and Letters on Natural Philosophy*. Cambridge, MA: Harvard University Press.

Collins, John David, Edward J Hall, and L. A Paul. 2004. *Causation and Counterfactuals*. Cambridge, MA: MIT Press.

Cushing, James T. 1990. *Theory Construction and Selection in Modern Physics: The S Matrix*. 1. publ. Cambridge, UK: Cambridge University Press. https://opacplus.bsb-muenchen.de/metaopac/search?db=100&View=default&lokalkey=1466025.

Demopoulos, William, and Michael Friedman. 1985. Bertrand Russell's *The Analysis of Matter*: Its Historical Context and Contemporary Interest. *Philosophy of Science* 52 (4): 621–39.

Eames, Elizabeth R. 1989. Cause in the Later Russell. In *Rereading Russell: Essays in Bertrand Russell's Metaphysics and Epistemology*, edited by C. Anthony Anderson and C. Wade Savage, 264–80. Minneapolis: University of Minnesota Press.

Earman, John. 2006. The "Past Hypothesis": Not Even False. *Studies in the History and Philosophy of Modern Physics* 37 (3): 399–430.

2011. Sharpening the Electromagnetic Arrow(s) of Time. In *The Oxford Handbook of Philosophy of Time*, edited by Craig Callender. Oxford: Oxford University Press.

Eberhardt, Frederick, and Richard Scheines. 2007. Interventions and Causal Inference. *Philosophy of Science* 74 (5): 981–95.

Einstein, Albert. 1909a. Über die Entwicklung unserer Anschauung über das Wesen und die Konstitution der Strahlung. *Physikalische Zeitschrift* 10 (22): 817–25.

1909b. Zum gegenwärtigen Stand des Strahlungsproblems. *Physikalische Zeitschrift* 10 (6): 185–93.

Evans, Denis J., and Debra J. Searles. 1996. Causality, Response Theory, and the Second Law of Thermodynamics. *Physical Review E* 53 (6) (June 1): 5808–15.

Farr, Matt, and Alexander Reutlinger. 2013. A Relic of a Bygone Age? Causation, Time Symmetry and the Directionality Argument. *Erkenntnis* 78 (S2): 215–35.

Fechner, Gustav Theodor. 1864. *Ueber die Physikalische und philosophische Atom-enlehre*. H. Mendelssohn.

Feynman, Richard P. 1965. *The Character of Physical Law*. Cambridge, MA: MIT Press.

Field, Hartry. 2003. Causation in a Physical World. In *The Oxford Handbook of Metaphysics*, edited by Michael J. Loux and Dean W. Zimmerman, 435–60. Oxford: Oxford University Press.

Frigg, Roman. 2010. Models and Fiction. *Synthese* 172 (2): 251–68.

Frisch, Mathias. 1999. Van Fraassen's Dissolution of Putnam's Model-Theoretic Argument. *Philosophy of Science* 66 (1): 158–64.

    2005a. *Inconsistency, Asymmetry, and Non-Locality: A Philosophical Investigation of Classical Electrodynamics*. Oxford: Oxford University Press.

    2005b. Mechanisms, Principles, and Lorentz's Cautious Realism. *Studies in History and Philosophy of Science Part B* 36 (4): 659–79.

    2005c. Counterfactuals and the Past Hypothesis. *Philosophy of Science* 72 (5): 739–50.

    2007. Causation, Counterfactuals and Entropy. In *Russell's Republic: The Place of Causation in the Constitution of Reality*, edited by Huw Price and Richard Corry, 351–95. Oxford: Oxford University Press.

    2009. "The Most Sacred Tenet"? Causal Reasoning in Physics. *The British Journal for the Philosophy of Science* 60 (3): 459–74.

    2010a. Causal Models and the Asymmetry of State Preparation. In *EPSA: Philosophical Issues in the Sciences. Launch of the European Philosophy of Science Association*, edited by M. Rédei, M. Dorato, and M. Suárez. Berlin: Springer.

    2010b. Does a Low-Entropy Constraint Prevent Us from Influencing the Past? In *Time, Chance, and Reduction: Philosophical Aspects of Statistical Mechanics*, edited by Andreas Hüttemann and Gerhard Ernst, 13–33. Cambridge, UK: Cambridge University Press.

    2011a. From Arbuthnot to Boltzmann: The Past Hypothesis, the Best System, and the Special Sciences. *Philosophy of Science* 78 (5): 1001–11.

    2011b. Principle or Constructive Relativity. *Studies in History and Philosophy of Science Part B* 42 (3): 176–83.

    2012. No Place for Causes? Causal Skepticism in Physics. *European Journal for Philosophy of Science* 2 (3): 313–36. doi:10.1007/s13194-011-0044-4.

    2013. Time and Causation, in *A Companion to the Philosophy of Time*, eds. Heather Dykes and Adrian Bardon, 282–300. Oxford: Wiley-Blackwell.

    2014. Why Physics Can't Explain Everything. In *Chance and Temporal Asymmetry*, edited by Alastair Wilson. Oxford: Oxford University Press.

Giere, Ronald N. 1988. *Explaining Science: A Cognitive Approach*. Chicago: University of Chicago Press.

Giere, Ronald N. 2006. *Scientific Perspectivism*. Chicago: University of Chicago Press.

Glauber, R., and V. I. Man'ko. 1984. Damping and Fluctuations in Systems of Coupled Quantum Oscillators. *Zhurnal Eksperimentalnoi i Teoreticheskoi Fiziki* 87 (September): 790–804.

Godfrey-Smith, Peter. 2009. Models and Fictions in Science. *Philosophical Studies* 143 (1): 101–16.

Goodman, Nelson. 1976. *Languages of Art: An Approach to a Theory of Symbols.* Indianapolis, IN: Bobbs-Merrill.

Griffiths, David J. 2004. *Introduction to Electrodynamics.* New Delhi: Prentice-Hall of India.

Halpern, Joseph Y., and Christopher Hitchcock. 2011. Actual Causation and the Art of Modeling. ArXiv e-print 1106.2652. http://arxiv.org/abs/1106.2652.

2014. Graded Causation and Defaults. ArXiv e-print 1309.1226. http://arxiv.org/abs/1309.1226.

Halpern, Joseph Y., and Judea Pearl. 2005. Causes and Explanations: A Structural-Model Approach. Part I: Causes. *British Journal for the Philosophy of Science* 56 (4): 843–87.

Hausman, D. M., and J. Woodward. 1999. Independence, Invariance and the Causal Markov Condition. *British Journal for the Philosophy of Science* 50 (4): 521–83.

Hausman, Daniel M. 1998. *Causal Asymmetries.* 1. publ. Cambridge Studies in Probability, Induction, and Decision Theory. Cambridge, UK: Cambridge University Press.

Hausman, Daniel M., and James Woodward. 2004. Modularity and the Causal Markov Condition: A Restatement. *The British Journal for the Philosophy of Science* 55 (1) (March 1): 147–61. doi:10.1093/bjps/55.1.147.

Healey, Richard. 1983. Temporal and Causal Asymmetry. In *Space, Time, and Causality*, edited by Richard Swinburne, 79–105. Dordrecht: Reidel.

1991. Review of *Asymmetries in Time: Problems in the Philosophy of Science* by Paul Horwich. *The Philosophical Review* 100 (1): 125–30.

Helmholtz, Hermann von. 1896. *Über das Ziel und die Fortschritte der Naturwissenschaft Eröffnungsrede für die Naturforscherversammlung zu Innsbruck 1869.* Heidelberg: Universitätsbibliothek der Universität Heidelberg.

Hertz, H. 1910. *Die Prinzipien der Mechanik* (2nd ed.). Leipzig: Johann Ambrosius Barth.

Hitchcock, Christopher. 2007. What Russell Got Right. In *Causation, Physics, and the Constitution of Reality: Russell's Republic Revisited*, edited by Huw Price and Richard Corry. Oxford: Oxford University Press.

Hoefer, Carl. 2004. Causality and Determinism: Tension, or Outright Conflict? *Revista Da Filosofia* 29 (2).

Holton, Richard. 2006. The Act of Choice. *Philosophers' Imprint* 6 (3).

Hoover, Kevin. 2001. *Causality in Macroeconomics.* Cambridge, UK: Cambridge University Press.

Horwich, Paul. 1987. *Asymmetries in Time: Problems in the Philosophy of Science.* Cambridge, MA: MIT Press.

Hume, David. 1975. Enquiry concerning Human Understanding. In *Enquiries concerning Human Understanding and concerning the Principles of Morals*, edited by L. A. Selby-Bigge, 3rd ed., revised by P. H. Nidditch. Oxford: Clarendon Press.

Hüttemann, Andreas. 2013. *Ursachen*. Berlin: De Gruyter.

Illari, Phyllis McKay, Federica Russo, and Jon Williamson. 2011. *Causality in the Sciences*. Oxford: Oxford University Press.

Jackson, John David. 1975. *Classical Electrodynamics* (2nd ed.). New York: Wiley.

Jackson, John David. 1999. *Classical Electrodynamics*. New York: Wiley.

Kinsler, Paul. 2011. How to Be Causal: Time, Spacetime, and Spectra. ArXiv e-print 1106.1792. http://arxiv.org/abs/1106.1792.

Kirchhoff, Gustav R. 1876. *Vorlesungen über mathematische Physik: Mechanik. I.* B. G. Teubner.

Ladyman, James, Otávio Bueno, Mauricio Suárez, and Bas van Fraassen. 2011. Scientific Representation: A Long Journey from Pragmatics to Pragmatics. *Metascience* 20 (3): 417–42.

Landau, Lev Davidovitch. 1975. *The Classical Theory of Fields*. Oxford: Butterworth-Heinemann.

Lewis, David. 1979. Counterfactual Dependence and Time's Arrow. *Noûs* 13 (4): 455–76.

1984. Putnam's Paradox. *Australasian Journal of Philosophy* 62 (3).

Lloyd, Seth. 1996. Causal Asymmetry from Statistics. In *Physical Origins of Time Asymmetry*, edited by J. J. Halliwell, J. Pérez-Mercader, and W. H. Zurek, 108–16. Cambridge, UK: Cambridge University Press.

Loewer, Barry. 2007. Counterfactuals and the Second Law. In *Causation, Physics, and the Constitution of Reality: Russell's Republic Revisited*, edited by Huw Price and Richard Corry, 293–326. New York: Oxford University Press.

2008. Why There Is Anything Except Physics. In *Being Reduced: New Essays on Reduction, Explanation, and Causation*, edited by Jakob Hohwy and Jesper Kallestrup (1st ed.). New York: Oxford University Press.

2012a. Two Accounts of Laws and Time. *Philosophical Studies* 160 (1): 115–37.

2012b. The Emergence of Time's Arrows and Special Science Laws from Physics. *Interface Focus* 2 (1) (February 6): 13–19.

Lorentz, Hendrik Antoon. 1904. Weiterbildung der Maxwellschen Theorie. Elektronentheorie. In *Encyclopädie der mathematischen Wissenschaften* 5 (2): 145–288.

Mach, Ernst. 1900. *Die Principien der Wärmelehre. Historisch-kritisch entwickelt.* Leipzig: Barth.

1901. *Die Mechanik in ihrer Entwickelung historisch-kritisch dargestellt.* Leipzig: Brockhaus.

1905. *Erkenntnis und Irrtum. Skizzen zur Psychologie der Forschung.* Leipzig: Barth.

Maudlin, Tim. 2007. *The Metaphysics within Physics*. New York: Oxford University Press.

Menzies, Peter, and Huw Price. 1993. Causation as a Secondary Quality. *British Journal for the Philosophy of Science* 44 (2): 187–203.

Mill, John Stuart. 1875. *A System of Logic*. London: Longmans, Green, Reader, and Dyer.

Newton-Smith, W. H. 1983. Temporal and Causal Asymmetry. In *Space, Time, and Causality*, edited by Richard Swinburne, 105–21. Dordrecht: Reidel.

Ney, Alyssa. 2009. Physical Causation and Difference-Making. *British Journal for the Philosophy of Science* 60 (4): 737–64.

Norton, John 2003. Causation as Folk Science. *Philosophers' Imprint* 3 (4): 1–22.

2007. Do the Causal Principles of Modern Physics Contradict Causal Anti-Fundamentalism? In *Thinking about Causes: From Greek Philosophy to Modern Physics*, edited by P. K. Machamer and G. Wolters, 222–34. Pittsburgh, PA: Pittsburgh University Press.

2009. Is There an Independent Principle of Causality in Physics? *British Journal for the Philosophy of Science* 60 (3): 475–86.

Nussenzveig, H. M. 1972. *Causality and Dispersion Relations*. New York: Academic Press.

Pearl, Judea. 2000. *Causality: Models, Reasoning, and Inference*. New York: Cambridge University Press.

2009. *Causality: Models, Reasoning, and Inference* (2nd ed.). New York: Cambridge University Press.

2011. The Structural Theory of Causation. In *Causality in the Sciences*, edited by Phyllis McKay Illari, Federica Russo, and Jon Williamson, 629–727. Oxford: Oxford University Press.

Penrose, O., and I. C. Percival. 1962. The Direction of Time. *Proceedings of the Physical Society* 79 (3): 605–16.

Pettersson, Thomas Sven, and P. Lefèvre. 1995. The Large Hadron Collider. *CERN Document Server*. http://cdsweb.cern.ch/record/291782.

Pippard, A. B. 1978. *The Physics of Vibration*. New York: Cambridge University Press.

Planck, Max. 1937. *Vom Wesen der Willensfreiheit*. Leipzig: Barth.

Popper, K. R. 1956a. The Arrow of Time. *Nature* 177 (4507) (March 17): 538.

1956b. Irreversibility and Mechanics. *Nature* 178 (4529) (August 18): 382.

Price, Huw. 1997. *Time's Arrow & Archimedes' Point: New Directions for the Physics of Time*. New York: Oxford University Press.

Price, Huw, Richard Corry, University of Sydney, and Centre for Time. 2007. *Causation, Physics, and the Constitution of Reality: Russell's Republic Revisited*. New York: Clarendon Press; Oxford University Press.

Price, Huw, and Brad Weslake. 2009. The Time-Asymmetry of Causation. In *The Oxford Handbook of Causation*, edited by Helen Beebee, Peter Menzies, and Christopher Hitchcock, 414–43. Oxford: Oxford University Press.

Putnam, Hilary. 1978. *Meaning and the Moral Sciences*. London: Routledge.

Reichenbach, Hans. 1956. *The Direction of Time*. Berkeley: University of California Press.

Reutlinger, Alexander. 2013. Can Interventionists Be Neo-Russellians? Interventionism, the Open Systems Argument, and the Arrow of Entropy. *International Studies in the Philosophy of Science* 27 (3): 275–95.

Rideout, D. P., and R. D. Sorkin. 1999. Classical Sequential Growth Dynamics for Causal Sets. *Physical Review D* 61 (2): 024002.

Ritz, Walther. 1908a. Recherches critiques sur l'électrodynamique générale. *Annales de chimie et de physique*.

1908b. Uber die Grundlagen der Elektrodynamik und die Theorie des schwarzen Strahlung. Physikalische Zeitschrift.

1909. Zum gegenwärtigen Stand des Strahlungsproblems. *Physikalische Zeitschrift* 10 (7): 224–5.

Ritz, Walther, and Albert Einstein. 1909. Zum gegenwärtigen Stand des Strahlungsproblems. *Physikalische Zeitschrift.*

Rohrlich, F. 2007. *Classical Charged Particles.* Hackensack, NJ: World Scientific.

Rohrlich, Fritz. 2006. Time in Classical Electrodynamics. *American Journal of Physics* 74 (4): 313. doi:10.1119/1.2178847.

Russell, Bertrand. 1912–3. On the Notion of Cause. *Proceedings of the Aristotelian Society* 13: 1–26.

1921. *The Analysis of Mind.* New York: Allen & Unwin; Macmillan.

1948. *Human Knowledge: Its Scope and Limits.* New York: Simon and Schuster.

1954. *The Analysis of Matter.* New York: Dover.

Scheibe, Erhard. 2006. *Die Philosophie der Physiker.* München: Beck. https://opacplus.bsb-muenchen.de/metaopac/search?db=100&View=default&lokalkey=9409045.

Schrödinger, Erwin. 1951. *Science and Humanism; Physics in Our Time.* Cambridge, UK: Cambridge University Press.

Schurz, Gerhard. 2001. Causal Asymmetry, Independent versus Dependent Variables, and the Direction of Time. In *Current Issues in Causation,* edited by Wolfgang Spohn, 47–67. Paderborn: Mentis Verlag.

Smith, Sheldon R. 2007. Causation and Its Relation to 'Causal Laws'. *British Journal for the Philosophy of Science* 58 (4): 659–88.

2013. Causation in Classical Mechanics. In *The Oxford Handbook of Philosophy of Physics,* edited by Robert Batterman, 107–40. Oxford: Oxford University Press.

Snider, Arthur David. 2006. *Partial Differential Equations: Sources and Solutions.* New York: Courier Dover.

Sober, Elliott. 1984. Common Cause Explanation. *Philosophy of Science* 51 (2): 212–41.

2001. Venetian Sea Levels, British Bread Prices, and the Principle of the Common Cause. *British Journal for the Philosophy of Science* 52 (2): 331–46.

Sommerfeld, Arnold. 1912. Die Greensche Funktion der Schwingungsgleichung. *Jahresbericht des Deutschen Mathematischen Veriens* 21: 309.

1968. *Gesammelte Werke Band 1.* Braunschweig: Friedrich Vieweg & Sohn.

Spirtes, Peter, Clark N. Glymour, and Richard Scheines. 2000. *Causation Prediction & Search 2e.* Cambridge, MA: MIT Press.

Steel, Daniel. 2005. Indeterminism and the Causal Markov Condition. *British Journal for the Philosophy of Science* 56 (1): 3–26.

2006. Homogeneity, Selection, and the Faithfulness Condition. *Minds and Machines* 16 (3): 303–17.

Steinhagen, R. J. 2007. LHC Beam Stability and Feedback Control. cdsweb.cern.ch/record/1054826/files/ab-2007–049.pdf.

Suárez, Mauricio. 2004. An Inferential Conception of Scientific Representation. *Philosophy of Science* 71 (5): 767–79.

2010. Scientific Representation. *Philosophy Compass* 5 (1): 91–101.

Suppes, Patrick. 1970. *A Probabilistic Theory of Causality*. Amsterdam: North-Holland.

Teller, Paul. 2001. Twilight of the Perfect Model Model. *Erkenntnis* 55 (3): 393–415.

Toll, John S. 1956. Causality and the Dispersion Relation: Logical Foundations. *Physical Review* 104 (6) (December 15): 1760–70. doi:10.1103/PhysRev.104.1760.

Uffink, Jos. 2006. Compendium to the Foundations of Classical Statistical Physics. In *Philosophy of Physics (Handbooks of the Philosophy of Science)*, edited by J. Butterfield and J. Earman. Amsterdam: Elsevier-North Holland.

van Fraassen, Bas C. 1980. *The Scientific Image*. Oxford: Oxford University Press.

1993. Armstrong, Cartwright, and Earman on Laws and Symmetry. *Philosophy and Phenomenological Research* 53: 431–44.

2008. *Scientific Representation: Paradoxes of Perspective*. Vol. 70. Oxford: Oxford University Press.

Weyl, Hermann. 1989. *The Open World: Three Lectures on the Metaphysical Implications of Science*. Woodbridge, CT: Ox Bow Press.

Wheeler, John Archibald, and Richard Phillips Feynman. 1945. Interaction with the Absorber as the Mechanism of Radiation. *Reviews of Modern Physics* 17 (2–3) (April 1): 157–81.

Williams, Bernard. 1978. *Descartes: The Project of Pure Enquiry*. New York: Penguin Books.

Winsberg, Eric. 2004. Can Conditioning on the "Past Hypothesis" Militate against the Reversibility Objection? *Philosophy of Science* 71:489–504.

Woodward, James. 2003. *Making Things Happen: A Theory of Causal Explanation*. Oxford: Oxford University Press.

2007. Causation with a Human Face. In *Causation, Physics, and the Constitution of Reality: Russell's Republic Revisited*, edited by Huw Price and Richard Corry. Oxford: Oxford University Press.

2012. A Functional Account of Causation Or: A Defense of the Legitimacy of Causal Thinking by Reference to the Only Standard That Matters – Usefulness (as Opposed to Metaphysics or Agreement with Intuitive Judgment). *Presidential Address delivered at the 2012 Philosophy of Science Association Meeting, San Diego*. Unpublished.

Zeh, H. D. 2007. *The Physical Basis of the Direction of Time*. New York: Springer.

Index

Printed in the United States
By Bookmasters